AF564705

Flowering Plants of Thrissur Forests

Flowering Plants of Thrissur Forests

Ajay Mukherjee

RANDOM PUBLICATIONS
NEW DELHI (INDIA)

Flowering Plants of Thrissur Forests

ISBN 978-93-5111-828-2

Published in 2016 in India by

RANDOM PUBLICATIONS

4376-A/4B, Gali Murari Lal, Ansari Road
New Delhi-110 002
Phone : +9111-43580356, 011-23289044, 011-43142548
e-mail: sales@randompublications.com,
info@randompublications.com, randomexports@gmail.com

Type Setting by : Friends Media, Delhi-110089
Digitally Printed at: Replika Press Pvt. Ltd.

Preface

Thrissur is situated in south western India and is in the central part of Kerala state. The center of the city is the Swaraj Round and it lies around small hillock called Thekin Kaadu Maidan on which the famous Vadakkumnathan temple is located. The city enjoys a tropical climate, the monsoons start in June. The months of April-May can get pretty humid which is also one of the best time to visit the city for the famed Thrissur Pooram(temple festival). The best weather is from October to February.

Kerala obtain good rain and has a tropical climate which is very suitable for growing flowering plants. Due to the climate in Kerala, there are many types of flowers found abundantly and naturally. Some flowers are common whereas there are some which are rare. There are flowers grown in home gardens and some are found only in the forest areas.

The mountain ranges with thick evergreen forests afford ideal abode for various animals and game including diverse birds while the middle country with hills and low plateau, mostly cleared for cultivation and human habitation, still affords shelter and food for many of the smaller mammals, birds and reptiles and also many lower animals of diverse groups. The lowlands of the extreme west, bordering the coastline are dotted with backwaters and estuaries of rivers, all connected by an interesting system of canals forming a continuous waterway. Its waters abound in fish and afford feeding ground for many water birds, local and migrant, while the plains have rich fauna representing all groups. Among the mammals the Primates are represented by the langurs and monkeys. Coconut palm and paddy are mainly cultivated in the lowlands.

The chief forest produce is timber. The principal local markets for timber are Cochin, Ernakulam and Thrissur. A large quantity of timber is transported to Coimbatore and Pollachi. High girth rosewood is exported to foreign countries. Other hard wood species which command a steady market are Irul, Pullamaruthu, Koramaruthu, Venga, Venteak, Pongu, Agil etc. Minor forest products are also abundant in the district. Mattaipal Karuvelampatta, Marotti, Poovam, Zamalporia, Kanjiram, Elavarngam are some of them. Kerala Forest Research Institute (KFRI) at Peechi is an institution which conducts ecological and Forests Development Research studies. The Institute has a very good nursery of medical plants.

The present work embodies the result of more than a decade's intensive floristic studies in one of the floristically very rich areas on the Western Ghats that is Thrissur (Kerala), the area of the present study have remained rather under-explored.

– Author

Contents

1

Introduction

FLOWERING PLANTS OF KERALA

Kerala with good rains and tropical climate is ideal for flowering plants to grow and profusely bloom. Orchids, Anthurium and Roses stand in front of the flowering plant category and some less prominent ones but more enchanting plants those are grown all over Kerala.

AAMBAL OR WATER-LILY

Scientific Name: *Nymphaea nouchali*
Common Name: Aambal or Water-lily
Family : *Nymphaeaceae*

The plant is known by many names like water lily, star-lotus, *Nymphaea stellata*etc. It is an aquatic plant with stem submerged in the ground and long petiole holding flat round leaves above water level. Water-lily flowers are brightly colored from red to blue; many petalled and float over water surface. Hindu myths have a prominent place for lotus flower and Lord Vishnu is supposed to be physically associated to lotus flower and all temples use lotus for rituals.

Fig. WATER LILLY (AAMBAL)

Water lillies are plants that grow in still or slowlt moving water. They like ponds, streams , and the edges of lakes in tropical and mild areas. Their floating leaves are often called lilly pads. Water lillies grow from the muddy bottom of a body of water. Thick underwater stems are buried in the mud.

The hairy water lily is an aquatic plant having erect perennial rhizomes or rootstocks that anchor it to the mud in the bottom. The rhizomes produce slender stolons.Its leave blades are round above the water and heart-shaped below 15–26(–50) cm, papery, abaxially densely pubescent. Some of the leaves that emerge rise slightly above the water held by their stem in lotus fashion, but most of them just float on the surface. The floating leaves have untie edges that make a crenellate effect. The water lily is also commercialized as an aquarium plant. The underwater leaves of this species have a handsome appearance that is appreciated by aquarists who often remove the floating leaves to keep it as a fully sub aquatic plant. The flowers are quite large, about 15 cm in diameter when fully open. They tend to close during the daytime and open wide at night. Their color varies from white to pink, mauve or purple depending from the variety or hybrid.

NYMPHAEA PUBESCENS

Fig. *Family:* NYMPHAEACEAE

Malayalam Name(s): Ambal, Neerambal, Periambal

English name(s): Hairy waterlily, Night lotus

Description: Perennial aquatic herbs; rhizomes ovoid, ca 3 x 2 cm. Leaves alternate, peltate or subpeltate, orbicular, hastate, sagittate or cordate at base, repand-dentate at margin, subacute, obtuse or rounded at apex, 15-40 x 12-35 cm, finely punctate and glabrous above and purplish green beneath; petioles smooth, ca 5 mm thick, purplish. Flowers 8-15 cm across, opening at night, closing from noon to evening; peduncles ca 15 mm thick. Sepals 4, ovate-lanceolate or elliptic-oblong, subacute at apex, 2.5-8 x 1-3 cm, puberulous, green, 5-9 ribbed outside, white and glabrous inside. Petals 10-25, oblanceolate, obtuse or acute at apex, 2-7 x 1-3 cm, white rose to pale pink. Stamens 25-70, pale yellow; filaments 1.5-3.5 cm long, dilated at base, yellow. Ovary 12-22 loculed; stigmatic appendages 5-10 mm long, incurved, closing over stigmas, yellow. Fruits globose, 3-4 cm, fleshy with remnants of stamens and appendaged stigmatic rays; seeds ellipsoid, ca 1.5 x 1 mm, longitudinally marked with rows of irregular papillae.

Habit: Herb

Flowering & Fruiting: November-December

District(s): Palakkad, Alappuzha, Kannur, Kollam, Malappuram, Kozhikkode, Wayanad, Thrissur, Ernakulam

Medicinal: Yes

Habitat: Fresh water pools, lakes and flooded paddy fields

Nymphaea nouchali

Nymphaea nouchali, often known by its synonym *Nymphaea stellata*, or by common names blue lotus, star lotus, red and blue water lily, blue star water lily is a water lily of genus *Nymphaea*. It is native to southern and eastern parts of Asia, and is thenational flower of Sri Lanka and of Bangladesh. This species is sometimes considered to include the blue Egyptian lotus *Nymphaea caerulea*. In the past there has been taxonomic confusion with the name *Nymphaea nouchali* incorrectly applied to *Nymphaea pubescens*.

Distribution and Habitat

Fig. The blue-flowered *Nymphaea nouchali* or *Nil Mânel*.

Fig. Fuchsia-colored *Nymphaea nouchali* in Hyderabad, Andhra Pradesh.

This aquatic plant is native in a broad region from Afghanistan, the Indian Subcontinent, toTaiwan, southeast Asia, and Australia. It has been long valued as a garden flower in Thailandand Myanmar to decorate ponds and gardens. In its natural state *N. nouchali* is found in static or slow-flowing aquatic habitats of little to moderate depth.

Description

Fig. Water lily in Thiruvananthapuram

Nymphaea nouchali is a day-blooming nonviviparous plant with submerged roots and stems. Part of the leaves are submerged, while others rise slightly above the surface. The leaves are round and green on top; they usually have a darker underside. The floating leaves have undulating edges that give them a crenellate appearance. Their size is about 20–23 cm and their spread is 0.9 to 1.8 m. This water lily has a beautiful flower which is usually violet blue in color with reddish edges. Some varieties have white, purple, mauve or fuchsia-colored flowers, hence its name red and blue water lily. The flower has 4-5 sepals and 13-15 petals that have an angular appearance making the flower look star-shaped from above. The cup-like calyx has a diameter of 11–14 cm.

Symbolism

N. nouchali is the national flower of Bangladesh. A pale blue-flowered *N. nouchali* is the national flower of Sri Lanka

In Sri Lanka this plant usually grows in buffalo ponds and natural wetlands. Its beautiful aquatic flower has been mentioned in Sanskrit, Pali and Sinhala literary works since ancient times under the names *kuvalaya*, *indhîwara*, *niluppala*, *nilothpala* and *nilupul* as a symbol of virtue, discipline and purity. Buddhist lore in Sri Lanka claims that this flower was one of the 108 auspicious signs found on Prince Siddhartha's footprint. It is said that when Buddha died, lotus flowers blossomed everywhere he had walked in his lifetime.

N. nouchali might have been one of the plants eaten by the Lotophagi of Homer's Odyssey.

Uses

N. nouchali is used as an ornamental plant because of its spectacular flowers. It is also popular as an aquarium plant under the name "Dwarf Lily" or "Dwarf Red Lily". Sometimes it is grown for its flowers, while other aquarists prefer to trim the lily pads, and just have the underwater foliage.

Nymphaea nouchali is considered a medicinal plant in Indian Ayurvedic medicine under the name *Ambal*; it was mainly used to treat indigestion.

Like all waterlilies or lotuses, its tubers and rhizomes can be used as food items; they are eaten usually boiled or roasted. In the case of*N. nouchali*, its tender leaves and flower peduncles are also valued as food.

The dried plant is collected from ponds, tanks and marshes during the dry season and used in India as animal forage.

ARALI, NERIUM PLANT, DOG-BANE, SOUTH-SEA ROSE, ADALPHA

Scientific Name: *Nerium indicum*

Common Name: Arali, Nerium plant, Dog-bane, South-sea rose, Adalpha

Family: *Apocyanaceae*

The plant is a popular shrub of Kerala garden grown for their extremely beautiful flowers. They grow as herbs, shrubs, trees and even as climbers. Their leaves are simple and alternate and flowers in beautiful inflorescence. Production of profuse sap is a specialty of the members of this family. This sap though poisonous (causes nausea and brain-damage if injested) is of much medicinal value (used for the treatment of scabies).

Nerium

Nerium oleander is an evergreen shrub or small tree in the dogbane family Apocynaceae, toxic in all its parts. It is the only species currently classified in the genus *Nerium*. It is most commonly known as oleander, from its superficial resemblance to the unrelated olive *Olea*. It is so widely cultivated that no precise region of origin has been identified, though southwest Asia has been suggested. The ancient city of Volubilis in Morocco may have taken its name from the Berber name *oualilt* for the flower. Oleander is one of the most poisonous of commonly grown garden plants.

Description

Fig. A seed capsule spreading seeds

Oleander grows to 2–6 m (6.6–19.7 ft) tall, with erect stems that splay outward as they mature; first-year stems have a glaucous bloom, while mature stems have a grayish bark. The leaves are in pairs or whorls of three, thick and leathery, dark-green, narrow lanceolate, 5–21 cm (2.0–8.3 in) long and 1–3.5 cm (0.39–1.38 in) broad, and with an entire margin. The flowers grow in clusters at the end of each branch; they are white, pink to red, 2.5–5 cm (0.98–1.97 in) diameter, with a deeply 5-lobed fringed corolla round the central corolla tube. They are often, but not always, sweet-scented. The fruit is a long narrow capsule 5–23 cm (2.0–9.1 in) long, which splits open at maturity to release numerous downy seeds.

Habitat and Range

Fig. Oleander shrub, Morocco

N. oleander is either native or naturalized to a broad area from Mauritania, Morocco, and Portugal eastward through the Mediterranean region and the Sahara (where it is only found sporadically), to the Arabian peninsula, southern Asia, and as far East as Yunnan in southern parts of China. It typically occurs around dry stream beds. *Nerium oleander* is planted in many subtropical and tropical areas of the world. On the East Coast of the US, it grows as far north as Washington DC, while in California and Texas it is naturalized as a median strip planting. In Sri Lanka this plant is called Kaneru , grown as an ornamental in gardens.

Ecology

Some invertebrates are known to be unaffected by oleander toxins, and feed on the plants. Caterpillars of the polka-dot wasp moth (*Syntomeida epilais*) feed specifically on oleanders and survive by eating only the pulp surrounding the leaf-veins, avoiding the fibers. Larvae of the common crow butterfly (*Euploea core*) also feed on oleanders, and they retain or modify toxins, making them unpalatable to would-be predators such as birds, but not to other invertebrates such as spiders and wasps.

The flowers require insect visits to set seed, and seem to be pollinated through a deception mechanism. The showy corolla acts as a potent advertisement to attract pollinators from a distance, but the flowers are nectarless and offer no reward to their visitors. They therefore receive very few visits, as typical of many rewardless flower species. Fears of honey contamination with toxic oleander nectar are therefore unsubstantiated.

Ornamental Gardening

Oleander grows well in warm subtropical regions where it is extensively used as an ornamental plant in landscapes, in parks, and along roadsides. It is drought-tolerant and will tolerate occasional light frost down to “10 °C (14 °F). It is commonly used in landscapingfreeway medians in California, Texas, and other mild-winter states in the Continental United States because it is upright in habit and easily maintained.

Its toxicity renders it deer-resistant. It is tolerant of poor soils and drought. Oleander can also be grown in cooler climates in greenhouses and conservatories, or as indoor plants that can be kept outside in the summer. Oleander flowers are showy and fragrant and are grown for these reasons. Over 400 cultivars have been named, with several additional flower colours not found in wild plants having been selected, including red, purple, pink, and orange; white and a variety of pinks are the most common. Many cultivars also have double flowers. Young plants grow best in spaces where they do not have to compete with other plants for nutrients. Thailand has produced miniature sized Nerium plants to use in ornamental gardening.

Therapeutic Efficacy

Drugs derived from *N. oleander* have been investigated as a treatment for cancer, unsuccessfully. According to the American Cancer Society, the trials conducted so far have produced no evidence of benefit, even though they did cause adverse side effects.

Toxicity

Oleander has historically been considered a poisonous plant because some of its compounds may exhibit toxicity, especially to animals, when consumed in large amounts. Among these compounds are oleandrin and oleandrigenin, known as cardiac glycosides, which are known to have a narrow therapeutic index and can be toxic when ingested.

Toxicity studies of animals administered oleander extract concluded that rodents and birds were observed to be relatively insensitive to oleander cardiac glycosides. Other mammals, however, such as dogs and humans, are relatively sensitive to the effects of cardiac glycosides and the clinical manifestations of "glycoside intoxication".

However, despite the common "poisonous" designation of this plant, very few toxic events in humans have been reported. According to the Toxic Exposure Surveillance System, in 2002, 847 human exposures to oleander were reported to poison centers in the United States. Despite this exposure level, from 1985 through 2005, only three deaths were reported. One cited death was apparently due to the ingestion of oleander leaves by a diabetic man. His blood indicated a total blood concentration of cardiac glycosides of about 20 ìg/l, which is well above the reported fatal level. Another study reported on the death of a woman who self-administered "an undefined oleander extract" both orally and rectally and her oleandrin tissue levels were 10 to 39 ìg/g, which were in the high range of reported levels at autopsy. And finally, one study reported the death of a woman who ingested oleander 'tea'. Few other details were provided.

In contrast to consumption of these undefined oleander-derived materials, no toxicity or deaths were reported from topical administration or contact with *N. oleander* or specific products derived from them. In reviewing oleander toxicity, Lanford and Boor concluded that, except for children who might be at greater risk, "the human mortality associated with oleander ingestion is generally very low, even in cases of moderate intentional consumption (suicide attempts)".

Toxicity studies conducted in dogs and rodents administered oleander extracts by intramuscular injection indicated that, on an equivalent weight basis, doses of an oleander extract with glycosides 10 times those likely to be administered therapeutically to humans are still safe and without any "severe toxicity observed".

In South Indian states such as Tamil Nadu and in Sri Lanka the seeds of related plant with similar local name (Kaneru(S)) Cascabela thevetia produce a poisonous plum with big seeds. As these seeds contain cardenolides, swallowing them is one of the preferred methods for suicides in villages.

Effects of Poisoning

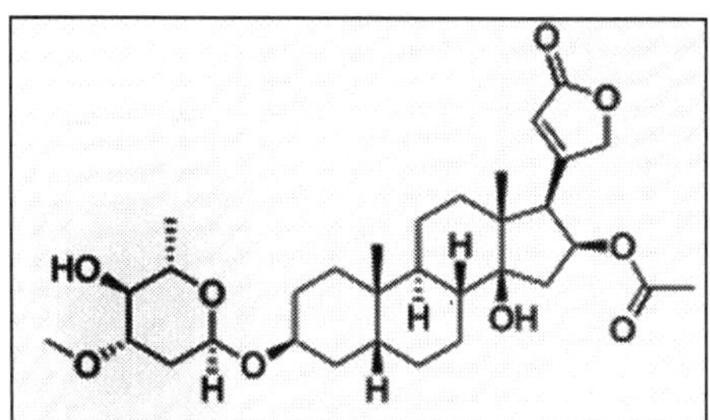

Fig. Oleandrin, one of the toxins present in oleander

Ingestion of this plant can affect the gastrointestinal system, the heart, and the central nervous system. The gastrointestinal effects can consist of nausea and vomiting, excess salivation, abdominal pain, diarrhea that may contain blood, and especially in horses, colic. Cardiac reactions consist of irregular heart rate, sometimes characterized by a racing heart at first that then slows to below normal further along in the reaction. Extremities may become pale and cold due to poor or irregular circulation. The effect on the central nervous system may show itself in symptoms such as drowsiness, tremors or shaking of the muscles, seizures, collapse, and even coma that can lead to death.

Oleander sap can cause skin irritations, severe eye inflammation and irritation, and allergic reactions characterized by dermatitis.

Treatment

Poisoning and reactions to oleander plants are evident quickly, requiring immediate medical care in suspected or known poisonings of both humans and animals. Induced vomiting and gastric lavage are protective measures to reduce absorption of the toxic compounds. Charcoal may also be administered to help absorb any remaining toxins. Further medical attention may be required depending on the severity of the poisoning and symptoms. Temporary cardiac pacing will be required in many cases (usually for a few days) till the toxin is excreted. Digoxin immune fab is the best way to cure an oleander poisoning if inducing vomiting has no or minimal success, although it is usually used only for life-threatening conditions due to side effects.

Drying of plant materials does not eliminate the toxins. It is also hazardous for animals such as sheep, horses, cattle, and other grazing animals, with as little as 100 g being enough to kill an adult horse. Plant clippings are especially dangerous to horses, as they are sweet. In July 2009, several horses were poisoned in this manner from the leaves of the plant. Symptoms of a poisoned horse include severe diarrhea and abnormal heartbeat. There is a wide range of toxins and secondary compounds within oleander, and care should be taken around this plant due to its toxic nature. Different names for oleander are used around the world in different locations, so, when encountering a plant with this appearance, regardless of the name used for it, one should exercise great care and caution to avoid ingestion of any part of the plant, including its sap and dried leaves or twigs. The dried or fresh branches should not be used for spearing food, for preparing a cooking fire, or as a food skewer. Many of the oleander relatives, such as the desert rose (*Adenium obesum*) found in East Africa, have similar leaves and flowers and are equally toxic.

Folklore

The alleged toxicity of the plant makes it the center of an urban legend documented on several continents and over more than a century. Often told as a true and local event, typically an entire family, or in other tellings a group of scouts, succumbs after consuming hot dogs or other food roasted over a campfire using oleander sticks.

Garden history

In his book *Enquiries into Plants* of *circa* 300 BC, Theophrastus described (among plants that affect the mind) a shrub he called*onotheras*, which modern editors render oleander; "the root of *onotheras* [oleander] administered in wine", he alleges, "makes the temper gentler and more cheerful".

The plant has a leaf like that of the almond, but smaller, and the flower is red like a rose. The plant itself (which loves hilly country) forms a large bush; the root is red and large, and, if this is dried, it gives off a fragrance like wine.

In another mention, of "wild bay" (*Daphne agria*), Theophrastus appears to intend the same shrub. Willa Cather, in her book *The Song of the Lark*, mentions oleander in this passage:

This morning Thea saw to her delight that the two oleander trees, one white and one red, had been brought up from their winter quarters in the cellar. There is hardly a German family in the most arid parts of Utah, New Mexico, Arizona, but has its oleander trees. However loutish the American-born sons of the family may be, there was never one who refused to give his muscle to the back-breaking task of getting those tubbed trees down into the cellar in the fall and up into the sunlight in the spring. They may strive to avert the day, but they grapple with the tub at last.

Oleander is the official flower of the city of Hiroshima, having been the first to bloom following the atomic bombing of the city in 1945.

It is the provincial flower of Sindh province.

Cultural Significance

- In the NBC series *Grimm*, the poison Spirit Oleander is used to kill a Rißfleisch (a tiger-like Wesen, or creature that a Grimm can differentiate from normal human beings) in combination with a specialized spear forged from pure silver.
- Professional Wrestler Alan Robert Rogowski, better known by his ring name, Ole Anderson adopted his moniker due to the toxicity of Nerium oleander.

In the 1946 Vincent Price movie Dragonwyck, based on Anya Seton's gothic novel, oleander is used as a poison by the villainous lead character.

DOGBANE

Scientific name: *Apocynum cannabinum L.*

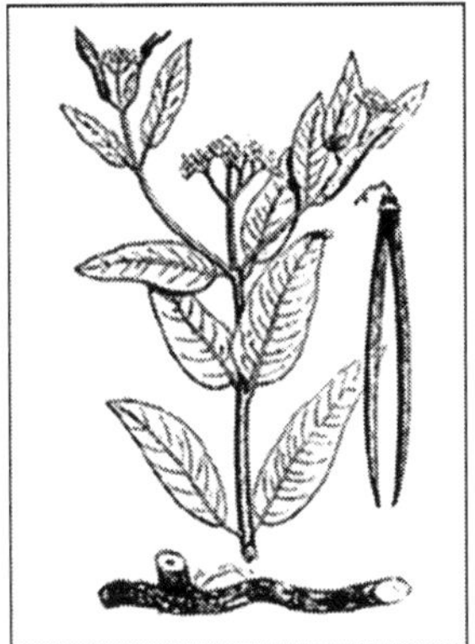

Fig. Kingdom: Plantae - Plants

Subkingdom: Tracheobionta - Vascular plants
Superdivision: Spermatophyta - Seed plants
Division: Magnoliophyta - Flowering plants
Class: Magnoliopsida - Dicotyledons
Subclass: Asteridae -
Order: Gentianales -
Family: Apocynaceae - Dogbane family
Genus: Apocynum L. - dogbane
Species: Apocynum cannabinum L. - Indian hemp

The common name, Dogbane, refers to the plant's toxic nature, which has been described as "poisonous to dogs." Apocynum means "Away dog!" and cannabinum means "like hemp," in reference to the strong cordage that was made by weaving together the stem's long fibers.

Description: A perennial with opposite leaves that secretes a milky sap when bruised or broken, reaching 5-6 feet in height. Found throughout the United States.

Growth Habit: Forb/herb
Duration: Perennial
U.S. Nativity: Native

Other common names used for "Dogbane" in California: Hemp dogbane, Indian hemp, Indian hemp dogbane

Other common names throughout the United States: Black hemp, Black Indian hemp, Canadian hemp, American hemp, Amy-root, Bowmans root, Bitterroot, Indian-physic, Rheumatism weed, Milkweed, Wild cotton, Choctaw-root, and General Marion's Weed.

Leaves: Entire margins (meaning the leaf's edges are smooth, not notched or toothed), ovate or elliptic, 2-5 inches long, 0.5-1.5 inches wide, and arranged oppositely along the stem. Leaves have short petioles (stems) and are sparingly pubescent or lacking hairs beneath. The lower leaves have stems while the upper leaves may not. The leaves turn yellow in the fall, then drop off.

Stems: Lack hairs, often have a reddish-brown tint when mature, become woody at the base, and are much-branched in the upper portions of the plant.

Roots: These plants may be found growing as colonies due to a long horizontal rootstock that develops from an initial taproot.

Flowers: Small, white to greenish-white, and produced in terminal clusters (cymes). The flower size is 1/4 inch wide. Blooms first appear in late spring and continue into late summer. The flowers are borne in dense heads followed later by the slender, pointed pods which are about 4 inches in length.

Many small insect pollinators, such as wasps and flies, pollinate the flowers.

Fruit: Long (5 inches or more), narrow follicles produced in pairs (one from each ovary) that turn reddish-brown when mature and develop into two long pods containing numerous seed with tufts of silky white hairs at their ends.Identifying Characteristics: Stems and leaves secrete a milky sap when broken. Sprouts emerging from the underground horizontal rootstock may be confused with Common Milkweed (Asclepias syriaca) emerging shoots. But note that they are not related to milkweeds, despite the milky sap and the similar leaf shape and growth habit. The flower shape is quite unlike that of milkweed flowers and the leaves of hemp dogbane are much smaller than those of common milkweed. When mature, these native plants may be distinguished by the branching in the upper portions of the plant that occurs in hemp dogbane, and also the smaller size of hemp dogbane compared to Common milkweed.

Habitat and range: Widespread in temperate areas from New England to Florida, parts of the southeast, Texas to California and north to British Columbia. Dogbane is a native of this country and may be found in thickets and along the borders of odd fields throughout the United States. Hemp dogbane occurs under moist conditions. Grows in average, dry to medium wet, well-drained soil in full sun. Commonly is found in gravelly or sandy fields, riparian areas, in meadows, along creek beds, irrigation ditches, hillside seeps, and fence lines in cultivated pastures. New plants begin growing in late spring or early summer.

Elevation: Between 0 and 7,000 feet

Notes: Cymarin, a chemical found in the plant's roots, was used as a cardiac stimulant, and was listed until 1952 in the medicinal text United States Pharmacopoeia. The milky sap contains cardiac glycosides (a chemical compound derived from a simple sugar and is often of medicinal importance) that have physiologic actions similar to digitoxin.

The toxic sap deters herbivores from feeding on the plant. Normally, animals avoid hemp dogbane because of its bitter, sticky, milk-white juice. Sheep are more frequently affected than other animals, as they will eat large quantities of hemp dogbane leaves and tops if other forages are not available. This most often occurs when animals are turned onto harvested fields or on fall ranges when forage is scarce. Poisoning can also occur when livestock are trailed from summer to winter ranges and other forages are not available.

Although this plant is considered toxic to humans, the roots were commonly harvested in the 19th and early 20th centuries for a variety of folk medicine and medical purposes.

Uses: Hemp dogbane is known for its use as cordage. Before domestic cotton was introduced and cultivated in the Southwest, around 700 A.D., leaf and stem fibers, hair or wool from dogs or wild animals, bird feathers, animal skins, or human hair were used to create prehistoric cordage. Dogbane fibers have been found in some archeological sites thousands of years old.

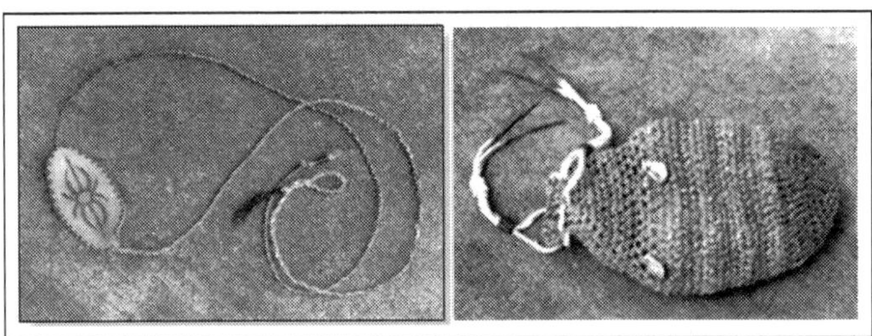

The native people from many nations in North America produced various useful items from the hemp fibers. They made cordage and thread from the plant with no other equipment than their hands and thighs. Fibers of the dogbane plant were rolled together to make a functional material stronger than cotton. The twine was excellent for making fishing lines and nets because it keeps its strength under water and does not shrink. It was also used in the manufacture of many other items, including deer and rabbit nets, slings for hunting small game, nooses for snaring grouse and other game birds, hide stretchers, bowstrings, moccasins, clothing, straps and tumplines, woven bedding for baby cradles, wheels used in a type of dart game, carrying nets, and cat-tail mats.

The stems were harvested late fall, after frosts have caused the sap to drop into the plant's perennial rootbase, and the leaves turn to yellow and fall from the stem. At this stage, the reddish-brown stalk becomes stiff and can easily be clipped from the rootbase flush to the ground. Dogbane stalks were also collected in winter, when the stalks are dry and brittle. Dry stalks are easiest to work, but if allowed to stand in the field through the winter and into spring, the fibers weaken. The harvested plants were bundled by the lower ends for carrying.

The cordage fibers are found in a layer between the thin outer skin and the woody, hollow center of the stem. Carefully scraping with a shell, an obsidian flake, chert spall, or deer rib removed the outer bark from the fibers. The top spreading branches were trimmed off and the stems were flattened between the fingers into four pieces. They were then split open from bottom to top. The length of brittle, woody core was broken into 2 inch pieces and pulled off the fibers by hand until all of the [* File contains invalid data | In-line.JPG *]woody material was removed. Then the process of rubbing and rolling the hemp between the hands helped clear away any other skin that still clinged to the fibers. Not everyone scraped the dogbane first. Paiute people and others simply split and broke the dry stalks without scraping, and removed all the outer bark by rubbing. Dogbane and other bast fibers can also be soaked (called "retting") to release the fibers. Retting was used when the bark adhered more tightly to the fibers.

The dogbane fibers were made into twine by twisting and rolling them with the hand on the bare thigh. The hands were kept damp to increase the friction. More stem fibers were joined together by splicing. The short end that needed splicing was overlapped with the new addition of fibers, then rolled together until they were intertwined. An average plant yields about 2 1/2 feet of fiber, but one fourth of this is lost in the splicing process. By splicing the stems together, a continuous length of twine could be produced. A strong rope could be made by plying several lengths of twine together. A good Indian hemp rope is said to have the equivalent strength of a modern hemp rope with a breaking point of several hundred pounds. The twine would keep for many years if stored in a dry place. In drier, more open areas, dogbane grows shorter (two or three feet), with more branches. In dense streambank thickets, the stems grow much longer and with fewer side branches, and are therefore more desirable for cordage. Different sizes of stalks and different growing conditions yield different colored fibers.

HEMP DOGBANE

Hemp dogbane (Apocynum cannabinum) is a member of the Dogbane family (Apocynaceae). A perennial broadleaved weed, it reproduces by seed and by its spreading root system.

Seedling Description

Hemp dogbane seedlings are erect and sturdy. The stem below the seed leaves (hypocotyl) is smooth, green, and often red at the base. Seed leaves (cotyledons) are oval, about ¼ inch (6 mm) long and half as wide, and green with a white midvein. The first true leaves are about ½ to 1 inch (12 to 25 mm) long, half as wide, bright green above and pale below. Leaf margins are smooth and the leaves themselves are rather thick for their size. Leaves usually have a short leaf stalk (petiole) but may be attached directly to the main stem.

Fig. Seedlings are smooth, erect, and sturdy.

Fig. Young plants will resprout if cut off at ground level.

Biology

Hemp dogbane is a member of the Dogbane family (Apocynaceae). A perennial broadleaved weed, it reproduces by seed and by its spreading root system. The seeds are extremely small and germinate best from a depth of 3/8 inch (9 mm), but in fine-textured soil may emerge from nearly 2 inches (5 cm) deep. Most seedlings emerge in spring, but shoots from the rhizomes may appear throughout the growing season, even into early fall. Seeds remain viable up to two years in soil. Germination may occur at any temperature between 50° and 104°F (10° and 40°C), but the optimum range for seedling development is between 68° and 77°F (20° and 25°C).

The plant emerges as a single stem but soon divides to form a bushy plant 1 to 5 feet (30 to !50 cm) tall. Stems are smooth, erect, and red to green, often turning entirely red in autumn. When broken, the stems release a milky white sap. Leaves are arranged in opposite pairs and have slightly pointed tips and a narrow oval shape. They grow 2 to 5 inches (5 to 13 cm) long and ¾ to 1¾ inches (2 to 4 cm) wide. Margins and upper leaf surfaces are smooth; lower surfaces may be slightly downy or entirely smooth. The upper surface is a dull dark-green and has a prominent network of white veins. Flowers appear in mid- to late summer. They form in round clusters called "cymes," about 2 inches across, in which the central flowers open first. Each urn-shaped flower is less than ¼ inch across and consists of five greenish-white petals.

Fig. Hemp dogbane commonly infests corn. Long seed pods are distinctive.

Some of the flowers become fertilized and in late summer develop into pairs of long, slender green pods called follicles. The pods are sharply pointed, 4 to 8 inches (10 to 20 cm) long and only about ¼ inch (3 mm) in diameter. When dry, they split lengthwise, each half twisting in a spiral and releasing the tiny flat seeds. Attached to each seed is a pappus, or tuft of silky hairs, that is easily carried away by the wind. Each plant produces an average of 20 pods, with about 200 seeds in each pod.

The weed's persistence is due to its perennial root system rather than its seeds. Ten-day-old seedlings already have perennial capabilities; they can resprout if cut off at ground level. Roots grow laterally and send up new shoots, so one hemp dogbane plant can quickly tum into a large patch. In fields under cultivation, tillage chops the roots into small pieces. A root segment less than one inch long with a single bud can produce a new plant.

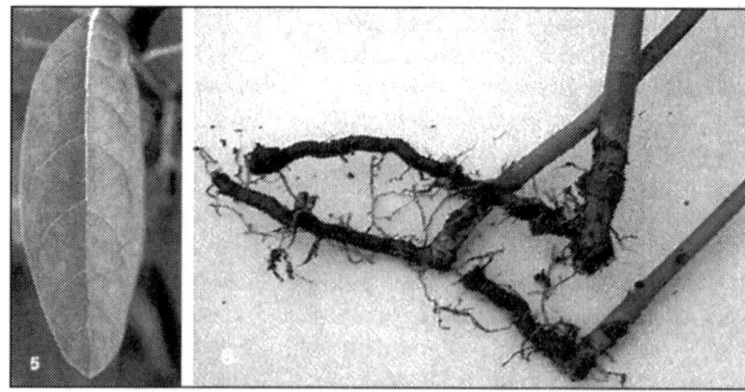

Fig. Leaves have a strong white midvein. Perennial, laterally spreading roots are difficult to control.

Similar Species

Common milkweed is often confused with hemp dogbane because the two weeds have similar leaves and sticky white sap. However, milkweed flowers are pink rather than green and have a more complicated structure than the simple five-petaled blossoms of hemp dogbane. Before flowers appear, the two plants can be distinguished by their growth habits. Hemp dogbane stems are multi-branched, forming a bushy plant. Milkweed has a single stalk, occasionally divided once or twice at the top.

When the plants are still very young, the leaves hold a clue to identification. Hemp dogbane leaves are smooth on both surfaces, whereas milkweed leaves usually have a soft downy covering on the lower surface.

Natural History

Hemp dogbane, a native of North America, grows throughout most of Canada and the United States, especially in the mid-Atlantic region. It grows in meadows, prairies, waste places and dumps, and along roadsides, streams, and woodland borders. It grows best in medium- to fine-textured soils, but it also does well in coarse soil if fertility and moisture are adequate.

Hemp dogbane is a serious problem in non-cultivated land, and thrives in moist fields. Crop reduction due to hemp dogbane infestation varies. Under irrigation, crop yield loss is slight, but non-irrigated sites in Nebraska showed a 15 percent reduction in corn, 32 percent loss in sorghum, and as much as a 37 percent loss in soybeans.

Much of the literature on hemp dogbane claims that it is poisonous to livestock, but these claims were based on an early investigation in which oleander (*Nerium oleander*) was mistaken for hemp dogbane. While the two plants belong to the same family and their leaves are similar, all parts of oleander are extremely poisonous to humans and all classes of livestock. As little as 0.005 percent of the animal's weight has proven fatal to horses; 12 ounces could be a lethal dose for a 1 ,500-pound horse. A human being can die from eating a single oleander leaf.

Hemp dogbane itself actually poses little danger. Animals find fresh hemp dogbane distasteful, but can eat it in hay without suffering ill effects.

Hemp dogbane extract has been used in medicinal preparations, and researchers have studied the effects of its compounds on animals. Two types of toxins—resins and glycosides—have been identified in hemp dogbane extracts. In concentrated form, the toxins cause cardiovascular problems, digestive disturbances, and death in dogs and cats.

Mature stems of hemp dogbane contain long, tough fibers which the Native American Indians used to make rope, bowstrings, nets, and thread. The Indians also used the milky juice to poison fish, and the root extract as a medicine. An archaeological dig in Utah uncovered a drawstring bag, from about 5000 B.C., made of Apocynum fibers. A natural dye can be made from hemp dogbane. When mixed with alum, the plant produces a dark tan color. Using copper as a mordant produces a black dye.The scientific name derives from three Greek words: *apo* (against), *kynakos* (dog), and*kannabin* (hemp). Hemp dogbane is also called Indian hemp, Indian physic, Choctaw root, bowman's root, and rheumatism weed.

Control

When conventional tillage was a standard farming practice, hemp dogbane was rarely a problem. In recent years, however, the use of selective herbicides in reduced-tillage farming has allowed hemp dogbane to form stands as large as 10 to 12 acres.

Prevention is an important part of hemp dogbane control. This weed is often introduced to clean fields when machinery carries root pieces from infested land. Establishment occurs quickly once the buds on these roots are allowed to sprout, so all machinery should be cleaned before moving on to the next field. Buying weed-free certified seed is strongly recommended. Fencerows and other noncultivated land around the farm are frequent sources of weed seed. Mowing these areas helps prevent reinfestation.

Control in corn is difficult because hemp dogbane seedlings emerge one to three weeks earlier than com seedlings, and shoots continue to emerge throughout summer. Herbicides applied as a preemergence or early postemergence treatment kill the early weed seedlings but have no effect on shoots that emerge later from root buds. Preharvest chemical control in corn would require an airplane or high-boy sprayer, so few farmers choose this treatment.

Crop rotation is one of the surest methods of subduing hemp dogbane. A rotation that includes com, small grain, and several years of hay provides effective competition against hemp dogbane. It also allows for chemical control in the fall after small grain harvest or in the last year of a hay crop.

Alfalfa is one of the best competitors against hemp dogbane. During the last year of the hay crop, a nonselective systemic herbicide should be applied three or four weeks *after* the last hay cutting. Allowing this length of time for regrowth insures that the dogbane will be large, actively growing, and able to translocate the chemical to its roots. Treatment should be at least one week before killing frost is expected, since the chemical works by moving through an actively growing plant.

For specific recommendations, consult your county Extension agent or the most recent *Weed Control Manual and Herbicide Guide*, available through Meister Publishing Company, 37841 Euclid Avenue, Willoughby, Ohio 44094. Follow label instructions for all herbicides and observe restrictions on grazing and harvesting procedures.

ARUMASAM, KRISHNA KIREEDAM OR PAGODA PLANT

Scientific Name: *Clerodentron paniculatum*
Common Name: Arumasam, Krishna kireedam or Pagoda plant
Family : *Verbenaceae*

The plant is a shrub that grows mostly in the wild but capable of excelling any garden-grown plant for its beautiful terminal inflorescence that lasts about six months. Orange-red- bell shaped flowers open row by row in the terminal cluster. Leaves are simple and dark green. The name attributed to it; 'Krishna kireedam means the crown of lord Krishna; as the inflorescence is so shaped; it is a pity that this plant is not given the adequate importance it deserves.

Clerodendrum paniculatum (PAGODA FLOWER)

The pagoda flower, so called because of its tall, pyramidal inflorescences, is one of the most spectacular *Clerodendrum* species.

Fig. *Clerodendrum paniculatum* growing in Singapore Botanic Gardens

Species information

Scientific name: *Clerodendrum paniculatum* L.
Common name: pagoda flower (English); pangil-pangil or panggil-panggil (Malay).
Conservation status: Least Concern according to IUCN Red List criteria.
Habitat: Various, from waste-ground to rainforest, often close to fresh water.
Key Uses: Ornamental. Medicinal.
Known hazards: None known.

Taxonomy

Class: Equisetopsida
Subclass: Magnoliidae
Superorder: Asteranae
Order: Lamiales

Family: Lamiaceae
Genus: *Clerodendrum*

About this species

Clerodendrum paniculatum was first described in 1767 by the 'father' of modern biological nomenclature – the Swedish botanist, Carl Linnaeus. The species epithet refers to the large 'paniculate' clusters of flowers (inflorescences), the feature which makes this such a visually-striking plant. The pagoda flower is commonly encountered in the Asian tropics, where it is popular as an ornamental and known for its medicinal uses.

Genus: Clerodendrum

Geography and Distribution

The pagoda flower is found throughout tropical and subtropical Asia, from Bangladesh to the Moluccas. It is widely cultivated and often establishes as a garden escapee in these regions, so that its original distribution is not entirely clear.

Description

The most distinctive features of *Clerodendrum paniculatum* are the large terminal inflorescences (thyrses, though often erroneously called panicles) up to 45 cm long, bearing numerous red-orange flowers. Each slender, tubular flower is 1.2–2 cm long with five small lobes, these usually being slightly paler than the tube. Butterflies are the main pollinators. They extend their long, thin proboscides into the flower tubes during which process pollen adheres to their bodies from the long-exserted stamens. The large, glossy, lobed leaves and fairly robust stems with an almost square cross-sectional form are also prominent characteristics of *C. paniculatum*. Their ability to produce root suckers allows pagoda flowers to spread vegetatively and they can form apparently clonal stands of several plants together.

Fruits and Seeds

Clerodendrum paniculatum frequently has a high percentage of aborted pollen grains and fruit does not appear to set among the populations observed in Java, New Guinea and Sri Lanka. Kew scientist Dr James Wearn has only seen two dried specimens in fruit (collected from Peninsular Malaysia) which, when dissected, had seeds (at least developing) within their fruits. No germination tests have been carried out at Kew to date.

Kew's work on this Species

The pagoda flower is one of about 150 species of *Clerodendrum*, a large genus which is native to Africa and Asia. *Clerodendrum* species can be found in a range of climatic conditions and habitats, from the temperate southern regions of China to the tropical heat of Borneo, and from primary rainforest to roadside scrub. Historically, there has been much confusion regarding *Clerodendrum* species concepts, such that herbarium specimens and cultivated plants were frequently misidentified. In order to address this problem, Dr James Wearn and Professor David Mabberley of Kew's Herbarium are currently undertaking research to review all of the species found in the Malesian region of southeast Asia. Clarification of this commonly-encountered genus is essential for taxonomists, ecologists and those undertaking practical conservation work.

Uses

The pagoda flower has a number of medicinal uses in Asia. In Malaysia an infusion is drunk as a purgative and is applied externally to distended stomachs. Various magical attributes have been recorded; indeed the Malay vernacular name pangil-pangil refers directly to the 'summoning' of spirits. *Clerodendrum paniculatum* is also supposed to confer protection from harm and is used as an elephant-medicine! Substances produced by several*Clerodendrum* species are undergoing more rigorous scientific trials in order to evaluate their medicinal potential. To date, results are promising, and antipyretic and anti-inflammatory properties have been verified, as well as antiviral activity.

Cultivation

The pagoda flower was taken into cultivation throughout Indomalesia many centuries ago, as it is easy to grow in warm, humid climates and produces large inflorescences nearly all year round. Flowers of cultivated plants are usually sterile and so do not produce fruits. During the eighteenth century, novel ornamental plants from 'the other side of the world' were in high demand in Europe and this species was one of the earliest to reach the foremost nurseries of the time, being introduced to Britain from Java in 1809 as a greenhouse plant. It is easily propagated vegetatively and strikes readily from cuttings.

An occasional, naturally-occurring colour form with pale lemon-yellow flowers and pedicels (previously called*Clerodendrum citrinum*) has been selected due to its ornamental appeal and is grown as *C. paniculatum* 'Alba'.

ASHOKAM, GALASOKA, ANGANAPRIYA OR ASHOKA TREE

Scientific Name: *Saraca ashoca*

Common Name: Ashokam, Galasoka, Anganapriya or Ashoka tree

Family : *Caesalpiniacea*

The plant is a holy tree for Indians; leaves flowers etc are extensively used in the preparations of Ayurvedic medicins. Ashoka plant bears beautiful scarlet colored flowers on all seasons and is widely grown in front of Indian houses. Ashoka gives dark shades in hot seasons for rest and recoup of the members of the family and hence called 'Angana-priya' (darling of the front-yard).

Saraca asoca

Saraca asoca (the ashoka tree; lit., "sorrow-less") is a plant belonging to the Caesalpinioideae subfamily of the legume family. It is an important tree in the cultural traditions of the Indian subcontinent and adjacent areas.

Description

Fig. Leaves and flowers in Kolkata, West Bengal, India

The ashoka is a rain-forest tree. Its original distribution was in the central areas of the Deccan plateau, as well as the middle section of the Western Ghats in the western coastal zone of the Indian subcontinent.

The ashoka is prized for its beautiful foliage and fragrant flowers. It is a handsome, small, erectevergreen tree, with deep green leaves growing in dense clusters. Its flowering season is around February to April. The ashoka flowers come in heavy, lush bunches. They are bright orange-yellow in color, turning red before wilting. Biologically, some of the flower's characteristics are very dry and abundant. This means that the flower is coated with a chemical on the outside.

As a wild tree, the ashoka is a vulnerable species. It is becoming rarer in its natural habitat, but isolated wild ashoka trees are still to be found in the foothills of the central and eastern Himalayas, in scattered locations of the northern plains of India as well as on the west coast of the subcontinent near Mumbai. There are a few varieties of the ashoka tree. One variety is larger and highly spreading. The columnar varieties are common in cultivation.

Mythology and tradition

The ashoka tree is considered sacred throughout the Indian subcontinent, especially in India,Nepal and Sri Lanka. This tree has many folklorical, religious and literary associations in the region. Highly valued as well for its handsome appearance and the color and abundance of itsflowers, the ashoka tree is often found in royal palace compounds and gardens as well as close to temples throughout India.

The ashoka tree is closely associated with the yakshi mythological beings. One of the recurring elements in Indian art, often found at gates of Buddhist and Hindu temples, is the sculpture of a yakshini with her foot on the trunk and her hands holding the branch of a flowering ashoka tree. As an artistic element, often the tree and the yakshi are subject to heavy stylization. Some authors hold that the young girl at the foot of this tree is based on an ancient tree deityrelated to fertility.

Yakshis under the ashoka tree were also important in early Buddhist monuments as a decorative element and are found in many ancient Buddhist archaeological sites. With the passing of the centuries the yakshi under the ashoka tree became a standard decorative element of Hindu Indian sculpture and was integrated into Indian temple architecture as salabhanjika, because there is often a confusion between the ashoka tree and the sal tree (*Shorea robusta*) in the ancient literature of the Indian subcontinent.

Fig. ashoka blossom

In Hinduism the ashoka is considered a sacred tree. Not counting a multitude of local traditions connected to it, the ashoka tree is worshipped in Chaitra, the first month of the Hindu calendar. It is also associated with Kamadeva, the Hindu god of love, who included an ashoka blossom among the five flowers in his quiver, where ashoka represent seductive hypnosis. Hence, the ashoka tree is often mentioned in classical Indian religious and amorous poetry, having at least 16 different names in Sanskrit referring to the tree or its flowers.

In *Mahâkâvya*, or Indian epic poetry, the ashoka tree is mentioned in the *Ramayana* in reference to the *Ashoka Vatika* (garden of ashoka trees) where Hanuman first meets Sita.

Other trees called 'ashoka tree'

A popular tree known as "false ashoka tree" or even as "ashoka tree", *Polyalthia longifolia*, is cultivated to resemble the growth pattern of erect *pillar-like* Mediterranean cypress trees. It is a popular park and garden plant, much used in landscaping on the Indian subcontinent. This tree can easily be distinguished by its compound leaves and very different flowers. Ashoka flowers are red (initially orange in color) while False Ashoka flowers are apple green in color. Ashoka fruits look like broad beans containing multiple seeds while false ashoka fruits are small, spherical and contain only one seed. Ashoka trees are small in height, while false ashoka is taller.

Fig. False ashoka

CHEMPAKAM OR JOY-PERFUME TREE

Scientific Name: *Michelia chempaca*
Common Name: Chempakam or Joy-perfume tree
Family : *Magnoliaceae*

The flowers are cup shaped and fleshy; as a primitive feature petals and sepals undistinguished (these parts are called tepals); flowers highly fragrant hence used as room fresheners; perfumes are extracted from its flowers. Indian ladies prefer to wear these flowers in hair and gents carry one in their pocket for the enchanting and long lasting perfume of this special flower.

CHEMBAKAM- WHITE PLUMERIA FLOWERS WITH EXOTIC FRAGRANCE|KERALA FLOWER PHOTOS

More to flower photography from Kerala.. This time presenting you photos of a bunch of pure white flowerbouquet of 'Chembakams' as they are named in Malayalam the local language of Kerala. Commonly theseflowers are termed as Frangipani or Plumerias.

Flowers with Exotic Fragrance-Pure White Plumeria or Frangipani Flower Facts and Photos-By pixelshots from Cochin

A bunch of white flower beauties who are not not just good to see but very sweet in nature too, provided your scale to goodness is fragrance and look, not strength. For me i will say a flower species i saw yet which has a fragrance beyond words. Something you have to really feel on, and its not your fault if you miss some moments unknowingly with her!..Everybody does if still alive and haven't lost senses..

Fig. Unity is Strength! Ofcourse lethal strength of flower Fragrance.Plumeria Flowers

Flowers and buds of Chembakam.Simple white flowers with white petals spreaded out from a yellow core. Have to say a masterpiece work contributed by God towards the flower kingdom. A nice pure white innocent member of the flower family. Coudn't get much information regarding the medicinal strength of the flower, but sure about is used in perfume industry. They can't help avoiding such a natural refreshing smell. Have to say may feel a little heady to some people, something like a stimulating concentrated fragrance.

Fig. Another photograph of the white bouquet of plumeria flowers.

This time the difference is used a cellphone camera for the snap unlike the digitalcam shoot done above. The plant is having slender branches and bleeds white fluid if you make cuts. It can grow like a tree and you can choose these plants as a part of your garden if you like fragranted garden air, particularly by sunrises as they keeps blooming with the sun..

Pulmeria Flower Facts and Information: Some details on Flowers Named Chembakam In Kerala/India: Talking some fact about the flowers, these are natives of tropical and subtropical Americas and now common in Asian continent. A plant which has nearly 10 species and a variety of colors among the flowers. They are usually shrubs or trees and often used in religious purposes. in Temples they are not used daily but during Balis.

Superstitions are their as these plants are places where mythical spirits and ghosts resides. An interesting thing is they keeps cheating the insects with their exotic hard fragrance. As they depends moths like insects for pollination, usually blooms in night and the insects keeps searching the flowers for nectar and the fact is that they have the fragrance only, not a single drop of honey. Pretty cool cheating huh.But came to know that the white sap of the plant is poisonous..

These flowers are photographed from the Subhash Park in Cochin/Kerala. Planted aside the cobble stoned sidewalk, they keeps smiling to people walking through. The park is an ideal place to get a refresh from the busy world just outside,particularly a nice place to spend your evenings with that orange tinted horizon and setting sun.. Talking about some photography, it comes easy to get the focus locked right on relatively bigger subjects like these flowers unlike macro. If you got a similarly colored background for the subject, often watched the camera brains in confusion and we have to sort it out with slight angle changes and refocusing. Anyway the more we experimenting the more we knows the photo gadget in our hands.

CHEMPARUTHI OR ROSE-MALLOW OR SORREL

Scientific Name: *Hibiscus rosasinensis*

Common Name: Chemparuthi or Rose-mallow or sorrel

Family : *Malvaceae*

The plant is a widely cultivated garden plant that bears long stalked single auxiliary (that arises from the axis of the leaves) flowers. A hibiscus flower is a typical one with sepals, calyx and petals and a long stalk formed of the fusion of male stamen bearing stalks that protects the long tube from the female part. Leaves are simple alternate dark green with toothed ridges. Plants are either herbs or shrubs and even as trees. At present in this group has gained much popularity due to a variety of colors and shapes being produced by way of biotechnology.

ROSE MALLOW

Fig. Landscape Designer

Rose mallow (*Hibiscus moscheutos*) is a large, fast-growing, cold hardy relative of hibiscus native to swamps and wetlands in the southeastern United States. Its enormous flowers and showy foliage are an asset to perennial borders in many parts of the country.

A Stunning Ornamental

This species of mallow, also known as swamp hibiscus, has become increasingly popular in recent years as a native alternative to the tropical hibiscus. Though it is found in wet soils in nature it grows quite well in ordinary garden soil as long as it has regular irrigation.

- Flowers - The flowers closely resemble that of tropical hibiscus, growing up to 12 inches in diameter on some cultivars. They are most often seen in either a crimson red or creamy white color though there are many other colors available.
- Foliage - The leaves are even bigger, up to eight inches long, and have a deep green color and interesting serrated pattern that make them just as showy as the flowers.
- Stalks - Rose mallow plants grow as a single branched stalk that emerges from the ground in spring.

The plants grow several inches a day until spreading their flowers in late summer at a height of six to eight feet though there are dwarf varieties available as well.

Gardening With Rose Mallow

Rose mallow grows in USDA zones 4-9 and prefers full sun and rich, moist soil. Long, hot summers are needed to get the plants to the flowering stage, making rose mallow unsuitable for high elevation areas, the Pacific Northwest and the cool coastal areas of California. It is one of the top perennials for the Deep South, however, where it thrives in the high humidity.

Availability

They are in the South and to a lesser extent throughout the eastern half of the country where they are best adapted. They can be purchased online if you don't find them in local nurseries.

Oxydendrum

Sourwood or sorrel tree, *Oxydendrum arboreum*, is the sole species in the genus *Oxydendrum*, in the family Ericaceae.

It is native to eastern North America, from southern Pennsylvania south to northwest Florida and west to southernIllinois; it is most common in the lower chain of the Appalachian Mountains. The tree is frequently seen as a component of oak-heath forests.

Fig. Foliage

Growth

Sourwood is a small tree or large shrub, growing to 10–20 m (33–66 ft) tall with a trunk up to 50 cm (20 in) diameter. Occasionally on extremely productive sites, this species can reach heights in excess of 30 meters and 60 cm diameter. The leaves are alternately arranged, deciduous, 8–20 cm (3.1–7.9 in) long and 4–9 cm (1.6–3.5 in) broad, with a finely serrated margin; they are dark green in summer, but turn vivid red in fall. The flowers are white, bell-shaped, 6–9 mm (1/4 to 1/3 inch) long, produced on 15–25 cm (5.9–9.8 in) long panicles. The fruit is a small woody capsule. The roots are shallow, and the tree grows best when there is little root competition; it also requires acidic soils for successful growth. The leaves can be chewed (but should not be swallowed) to help alleviate a dry-feeling mouth.

Description

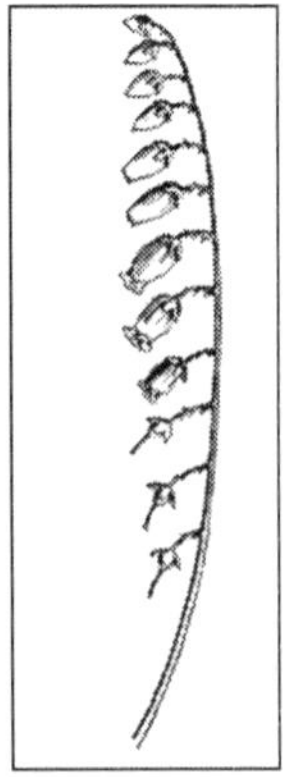

Fig. Raceme of flowers

The bark is gray with a reddish tinge, deeply furrowed and scaly. Branchlets at first are light yellow green, but later turn reddish brown. The wood is reddish brown, with paler sapwood; it is heavy, hard, and close-grained, and will take a high polish. Its specific gravity is 0.7458, with a density of 46.48 lb/cu ft.

The winter buds are axillary, minute, dark red, and partly immersed in the bark. Inner scales enlarge when spring growth begins. Leaves are alternate, four to seven inches long, 1.5 to 2.5 inches wide, oblong to ablanceolate, wedge-shaped at the base, serrate, and acute oracuminate. Leaf veins are Feather-veined, the midrib is conspicuous. They emerge from the bud revolute, bronze green and shining, and smooth; when full grown, they are dark green, shining above, and pale and glaucous below. In autumn, they turn bright scarlet. Petioles are long and slender, with stipules wanting. They are heavily laden with acid.

In June and July, perfect, cream-white flowers are borne in terminal panicles of secund racemes seven to eight inches long; rachis and short pedicels are downy. The calyx is five-parted and persistent; lobes are valvate in bud. The corolla is ovoid-cylindric, narrowed at the throat, cream-white, and five-toothed. The 10 stamens are inserted on the corolla; filaments are wider than the anthers; anthers are two-celled. The pistil is ovary superior, ovoid, and five-celled; the style is columnar; the stigma is simple; the disk is ten-toothed, and ovules are many. The fruit is a capsule, downy, five-valved, five-angled, and tipped by the persistent style; the pedicels are curving.

Cultivation and uses

The sourwood is perfectly hardy in the north and a worthy ornamental tree in lawns and parks. Its late bloom makes it desirable, and its autumnal coloring is particularly beautiful and brilliant. The leaves are heavily charged with acid, and to some extent have the poise of those of the peach. The leaves are also a laxative. It is renowned for nectar, and for the honey which is produced from it. Juice from its blooms is used to make sourwood jelly. The shoots were used by the Cherokee and the Catawba to make arrowshafts.

KASHITHETTI, SAVAM-NARI OR MADAGASCAR PERIWINKLE

Scientific Name: *Catharanthus roseus* (*Vinca rosea)*

Common Name: Kashithetti, Savam-nari or Madagascar periwinkle

Family : *Apocynaceae*

It is an herb that bears white or pink flowers on all seasons. Leaves are small, simple and green; fruits paired follicles (2 to 4 cm long). Periwinkle contains alkaloids of medicinal values and cultivated in large scale.

SHAVAM NARI POO (VINCA)

Shavamnari is a type of flower . It is also known as shavamnari poo , kashithumba (in malayalam) , vinca , vinca rose , lavendar vinca ,periwinckle flower , periwinckle vinca . Its Scientific name is Catharanthus Roseus. Its Very common in Kerala as a garden Plant. It is medicinal plant. It an also be poisonous if used inappropriately. Internally the leaves has been used for improved oxygen and blood flow , especially in cerebral veins . Externally a oil or oilment can be made or the leaves can be bruised and applied directly for dermatitis eczema ,acne , bleeding gums , nosebleeds , mouth ulcers, cancer.

CATHARANTHUS ROSEUS (MADAGASCAR PERIWINKLE)

Madagascar periwinkle is a popular ornamental plant found in gardens and homes across the world, and is used in the treatment of cancer.

Fig. *Catharanthus roseus* (Madagascar periwinkle)

Species information

Scientific name: *Catharanthus roseus* (L.) G.Don
Common name: Madagascar periwinkle, rosy periwinkle
Conservation status: Not yet assessed according to IUCN Red List criteria.
Habitat: On sand and limestone soils in woodland, forest, grassland, and disturbed areas.
Key Uses: Medicinal, ornamental.
Known hazards: As with other members of the Apocynaceae family, the sap is extremely toxic.

Taxonomy

Class: Equisetopsida
Subclass: Magnoliidae
Superorder: Asteranae
Order: Gentianales
Family: Apocynaceae
Genus: *Catharanthus*

About this species

Madagascar periwinkle is a popular ornamental plant found in gardens and homes across the warmer parts of the world. It is also known as the source of chemical compounds used in the treatment of cancer. Their discovery led to one of the most important medical breakthroughs of the twentieth century.

The flowers are adapted to pollination by a long-tongued insect, such as a moth or butterfly. This species is also able to self-pollinate. Self-compatibility and a relatively high tolerance to disturbance have enabled the plant to spread from cultivation and to become naturalised in many parts of the world. As a consequence, Madagascar periwinkle is sometimes considered to be an invasive weed, although it does not normally proliferate sufficiently to eliminate native vegetation. Its seeds have been seen to be distributed by ants.

Genus: Catharanthus

Geography and distribution

Madagascar periwinkle is indigenous to Madagascar, but is cultivated and naturalised throughout the tropics and parts of the subtropics.

Fig. *Catharanthus roseus* bush

Description

Overview: A tender, perennial plant which grows as a herb or subshrub sprawling along the ground or standing erect (30 cm to 1 m in height). Like many other plants in the Apocynaceae family, the sap is a milky latex.

Flowers: It has attractive white or pink flowers comprising five petals spreading from a long, tubular throat.

Leaves: The leathery, dark green leaves are arranged in opposite pairs.

Fruits: Each fruit is made up of two narrow, cylindrical follicles which house numerous grooved seeds.

Medicinal

In traditional medicine, Madagascar periwinkle has been used to treat a variety of ailments in Madagascar as well as in other parts of the world where the plant has naturalised. Whilst researching the anti-diabetic properties of the plant in the 1950s, scientists discovered the presence of several highly toxic alkaloids in its tissues. These alkaloids are now used in the treatment of a number of different types of cancer, with one derived compound, called vincristine, having been credited with raising the survival rate in childhood leukaemia from less than 10% in 1960 to over 90% today. Powerful medicinal plants such as the Madagascar periwinkle remind us of the need to conserve and study the increasingly threatened plant habitats of the world.

Ornamental

Madagascar periwinkle is grown as a bedding plant in tropical regions and cultivated indoors as a house plant in temperate areas.

Threats and Conservation

Madagascar's forests have been heavily impacted by human activity, but Madagascar periwinkle's ability to thrive in disturbed areas has enabled it to survive in its island home. Furthermore, it is widely established in the wild throughout tropical regions of the world, commonly cultivated in gardens and homes, and grown commercially for the pharmaceutical industry. Given the modern ubiquity of naturalised and cultivated populations of Madagascar periwinkle, direct conservation measures are a low priority for this species.

A rainforest Cure for Cancer?

Madagascar periwinkle was for many years grown simply as an attractive bedding plant in tropical areas. It comes in a range of pinks and reds that give rise to its other common name, rosy periwinkle. But today it has a more serious purpose; planted around the world during colonial times, it quickly became known at the same time as a useful folk medicine for diabetics. American and Canadian researchers during World War Two became aware that soldiers stationed in the Philippines were using it instead of insulin during shortages.

As a consequence, after the war, lab testing was done in earnest on Madagascar periwinkle. This revealed that the plant contained thirty alkaloids, strong plant chemicals that might be of use to humans. The leaves were found to contain two particularly valuable alkaloids, vinblastine and vincristine. These alkaloids work by disrupting part of cell division or 'mitosis', stopping the process when newly copied DNA is split into two identical parts to produce two identical new cells.

The drug company Eli Lilly tested the new chemicals on mice and found they helped to combat cancers, in particular those of the bone marrow such as childhood leukaemia and non-Hodgkin's Lymphoma. They developed an effective chemotherapy treatment, and today the prognosis for sufferers has changed from a one in ten chance of survival, to at least an eight out of ten chance of some remission.

Contracts and Collecting

Fig. *Catharanthus roseus* in the wild

Madagascar periwinkle shows how important it is to preserve areas of rich biodiversity, for there may be other 'miracle drugs' contained in plants just waiting to be discovered. But it also epitomises the issues involved in exploiting the riches of the plant world. In the 1960s, when these alkaloids were discovered, rich Western countries saw little need to reward the original homeland of the periwinkle, Madagascar, for the riches that had been earned from the plant.

Today at Kew we take a different view. We believe that Madagascar's people should have a say over how their plant resources are exploited: particularly where a plant is restricted to a single country. The Convention on Biological Diversity, of which the UK is a signatory, states that rights to exploit a discovery like that of the alkaloids in the periwinkle should be shared fairly.

In practice, this means that Kew has a legal team whose job is it to make sure that things are done equitably. Before a plant collecting team can even book flights, contracts are drawn up to ensure that the country being visited is happy with the arrangements being made. With such contracts in place, local people have much more incentive to look after their biodiversity. Now they know that if a profitable discovery is made, they will have a share in it.

Millennium Seed Bank: Seed storage

Kew's Millennium Seed Bank Partnership aims to save plant life worldwide, focusing on plants under threat and those of most use in the future. Seeds are dried, packaged and stored at a sub-zero temperature in our seed bank vault.

A collection of *Catharanthus roseus* seeds is held in Kew's Millennium Seed Bank based at Wakehurst in West Sussex.

Cultivation

Madagascar periwinkle is easy to cultivate, and can be propagated by seed or by cuttings, but is sensitive to over-watering. A tender plant, it does not withstand frosts and is best grown indoors in temperate climates. It thrives in hot and humid environments, in full sun or partial shade and flowers all year round in hot climates.

Catharanthus roseus is easily propagated by apical semi-ripe cuttings in light, free-draining compost. The best results are obtained when bottom heat and high humidity are provided. Propagation can also be carried out by seed, which should be maintained at 22-25°C and kept in the dark until the seeds germinate.

KADALADI, MEXICAN CREEPER, ROSA DE MONTANA, QUEEN'S WREATH OR CORAL VINE

Scientific Name: *Antigonon leptopus*

Common Name: Kadaladi, Mexican creeper, Rosa De Montana, Queen's wreath or Coral vine

Family : *Polygonaceae*

The plant is a tendril-climber that grows wild it is invasive and troublesome. Its profuse flowers that grow in beautiful inflorescences are capable to make it a garden plant blush. Its flexible stem is used to make baskets by the tribal people.

Antigonon leptopus

Antigonon leptopus, commonly known as Mexican creeper, coral vine, bee bush (in most Caribbean islands) or San Miguelito vine, is a species of flowering plant in the buckwheat family, Polygonaceae. It is a perennial that is native to Mexico. It is a vine with pink or white flowers (*Antigonon leptopus 'alba'*).

Invasive species

It is listed as a category II invasive exotic by the Florida's pest plant council.

Fig. Antigonon leptopus in China,from*Flower View*

Description

Antigonon leptopus is a fast-growing climbing vine that holds via tendrils, and is able to reach 25 ft or more in length. It has cordate (heart shaped), sometimes triangular leaves 2½ to 7½ cm long the flowers are borne in panicles, clusted along the rachis producing pink or white flowers from spring to autumn, it forms underground tubers and large rootstocks, it is a prolific seed producer, the seeds float on water, the fruit and seeds are eaten and spread by a wide range of animals such as pigs, raccoons and birds. The tubers will resprout if it is cut back or damaged by frost.

Fig. Coral Vine photographed at Jaipur, India.

Fig. This is a white colored flowering variety of coral vine at AC&RI, Killikulam, India

Uses

Antigonon leptopus was prepared for consumption by the aboriginal inhabitants of Baja California in a way reminiscent of popcorn. The seeds were toasted by placing them in a flat basket made of flexible twigs torn into several strips and woven to make a solid surface. On top of the seeds they would put live coals, and with both hands they would shake the basket so that the coals come up against the seeds, toasting them but not burning the basket. When the toasting is finished the burned out coals are removed and a major portion of the seeds are burst open exposing a white meal. Afterwards the seeds are separated from the husks from which they have come by dextrously tossing them into the air with the basket, just as wheat is winnowed in Spain. Thus cleaned they would grind and eat the prepared meal. They would also boil it and make fried cakes.

KANAKAMBARAM, ABULI OR FIRE-CRACKER PLANT

Scientific Name: *Crossandra infundibuliformis*

Common Name: Kanakambaram, Abuli or fire-cracker plant

Family : *Acanthaceae* or *Ruellia*

The plant is a very popular flowering plant as its brightly colored flowers (blue to golden yellow) are used by women to adorn their hair. This plant is a shrub with simple oval leaves and its ripe fruits bearing pods burst when come in to contact with water; hence a favorite game for rural children and gained it the name fire-cracker plant.

Crossandra infundibuliformis

Crossandra infundibuliformis (firecracker flower), is a species of flowering plant in the family Acanthaceae, native to southern India and Sri Lanka.

Description

Fig. Crossandra Infundibuliformis,Tumkur, India

It is an erect, evergreen subshrub growing to 1 m (3 ft 3 in) with glossy, wavy-margined leaves and fan-shaped flowers, which may appear at any time throughout the year. The flowers are unusually shaped with 3 to 5 asymmetrical petals. They grow from four-sided stalked spikes, and have a tube-like ¾ inch stalk. Flower colours range from the common orange to salmon-orange or apricot, coral to red, yellow and even turquoise.

Cultivation and uses

This plant requires a minimum temperature of 50 °F (10 °C), and in temperate regions is cultivated as a houseplant. It is usually grown in containers but can be attractive in beds as well. The flowers have no perfume but stay fresh for several days on the bush. A well-tended specimen will bloom continuously for years. It is propagated by seeds or cuttings.

Fig. Fire Cracker Flower - Tamil Nadu, South India

This plant has gained the Royal Horticultural Society's Award of Garden Merit. The tiny flowers are often strung together into strands, sometimes along with white jasmine flowers and therefore in great demand for making garlands which are offered to temple deities or used to adorn women's hair.

Fig. Fire Cracker Flower - Tamil Nadu, South India

Name

The common name "firecracker flower" refers to the seed pods, which are found after the flower has dried up, and tend to "explode" when near high humidity or rainfall. The "explosion" releases the seeds onto the ground, thereby creating new seedlings. It is popularly known as Kanakambaram in Tamil, Malayalam and Telugu and Kanakambara in Kannada. In Maharashtra it is known as Aboli.

FIRECRACKER PLANT

Russelia equisetiformis

Firecracker plant grows in a wild and wispy free-form with cascading fiery red blooms that attract hummingbirds and butterflies.

Fig. The tubular flowers, most common in red, resemble a fountain-like burst of fireworks that make this plant showy and appealing.

If ever there was a plant that epitomized the "right plant, right place" rule - this is it. Decide if you really have the space because a firecracker will not be tamed to stay small and compact. It looks terrible if you try to keep it that way.

Firecrackers need plenty of room for their signature arching form. But that's not to say you can't control one...just plan ahead so its size works in the area you want to plant.

This is not a good choice for the neat freak who wants a formal, manicured look - though the addition of one of these shrubs can soften and highlight an otherwise well-trimmed landscape. The fine-textured waterfall of stems with sparkling red flowers has more of a wildflower, cottage garden appeal.

We've included this plant in the Small Shrubs section for height - it can be kept about 3 feet tall. But an individual plant can spread to 5 feet wide and may overwhelm nearby smaller plants. Avoid placing anything directly in front of a firecracker plant. To best show off its beautiful form, use it at the end of a garden bed, around palm trees, or in front of taller, more upright shrubs such as hibiscus or firespike.

Plant specs

Firecrackers are moderate to fast growers to 3 feet tall and 3 to 5 feet wide. They prefer full to part sun but will grow in part shade (though you won't get as many flowers). These salt-tolerant shrubs do best in Zone 10, though you can plant in a container in Zone 9B and bring inside during cold weather.

Flowering takes place on and off all year, more during warmer months.

Plant care

Add top soil or organic peat humus to the hole when you plant. You can also add composted cow manure to the mix to enrich the soil around the root ball.

Cut back too tall or wayward shoots, if you like, but avoid a hard pruning of the entire plant - it may never grow out of it and recover its original beauty.

If the plant is damaged by cold, trim stems sparingly and let new growth emerge and cover the old (which you can remove later if you like). Try to leave some flowers on for the regular crowd of hummingbirds.

Fertilize 3 times a year - spring, summer and fall - with a controlled-release fertilizer.

Because this shrub covers a lot of area, controlled release can be used close to the plant without burning it. Supplement feedings, if you like, with liquid fertilizer to promote heavier bloom. A firecracker needs regular waterings to thrive and look its best.

If you have to hand water or you're on an irrigation system better timed for drought-tolerant shrubs, plant with water-retention crystals to keep the plant hydrated during dry spells.

Plant spacing

Place these shrubs about 3 to 4 feet apart. Come out from the house 3 feet or more.

Plant well away from walkways - preferably 3 to 4 feet - so you won't eventually be stepping on stems and flowers.

Fig. Utterly fabulous in containers or planters, firecrackers will drape beautifully over the sides.

KANIKKONNA OR GOLDEN-SHOWER TREE IS A SHRUB

Scientific Name: *Cassia fistula*
Common Name: Kanikkonna or golden-shower tree is a shrub
Family: *Fabaceae*

The plant is the State of Kerala's official flower, as the name denotes its flowers are golden-yellow and grown in long pendent inflorescence. The 'golden shower tree' produces flowers early and considered as a clarion-call of the arrival of Vishu or Vaishakh a festival of flowers (golden flowers of this plant is used to adorn the image of Lord Krishna on the day of Vishu and shown to children as the first sight of the day for Vishu). A fully blown golden shower tree is a treat to the eye of the beholder and the fallen petals on the ground gives it the name 'golden-shower' tree.

KONGINI OR BENDHI OR POT-MARIGOLD PLANT

Scientific Name: *Calendula officianalis*
Common Name: Kongini or bendhi or Pot-marigold plant
Family : *Austeraceae*

The plant is a weak stemmed annual herb with compound dark green leaves; it is widely grown for its aromatic and beautiful flowers (inflorescence with outer layer of petals forming the colorful part); aromatic compounds and dyes extracted from flowers. Marigold flowers are worn by women on auspicious occasions like marriage and celebrations.

KULAMARIYAN, MADHUMALTI OR RANGOON OR CHINESE CREEPER

Scientific Name: *Quisqualis indica*
Common Name: Kulamariyan, Madhumalti or Rangoon or Chinese creeper
Family : *Combretaceae*

The plant is an invasive, spiny creeper that grows wild and spread quick covering the whole area by producing roots and fresh spouts from creeping stem.

Its inflorescences are pendent, flowers five petalled, rich in honey with a tubular stalks; appear white when freshly blown and turns red by passing of the day. Leaves of this plant are used as anti-worm potion. This plant has some medicinal uses and its stem is used for making baskets by tribal people.

MANDARAM OR DWARF WHITE BAUHINIA

Scientific Name: *Bauhinia acuminata*
Common Name: Mandaram or Dwarf white bauhinia
Family : *Fabaceae*

It is a garden grown plant that grow only up two meters in height (certain varieties grow as trees bearing blue flowers and often grown on street-sides). Generally dwarf bauhinias bear white shiny flowers on all seasons; flowers are with five petals; leaves camel-foot shaped (hence it is known as camel-foot plant also) and broad, fruits are pods (15 cm in length and 1.5 cm broad; bear 6 to 10 seeds.

MUKKUTTI OR LIFE-PLANT

Scientific Name: *Biophytum sensitivum*
Common Name: Mukkutti or life-plant
Family : *Oxalidaceae*

A short herb with pinnate leaves (eight to seventeen in number) that radiates from the stem, flowers yellow; five petalled long stalked. *Biophytum* plant has many names like Aleluya in French, nilakurunji in Tamil, Lejjalu in Gujarathi etc. It collapses when stirred (hence the Hindi name Lajjalu meaning shy) it regains shape afterwards the commotion. Mukkutti is of high medicinal value and widely used in Ayurveda.

MULLA OR JASMINE

Scientific Name: *Jasminum grandiflorum*
Common Name: Mulla or Jasmine
Family : *Oleaceae*

This is one of the most popular flowers for its enchanting fragrance; flowers are white, five petalled, with tubular stalk; leaves generally simple and dark-green. About 200 species exist in this group.*Jasminum sambac* has many petals (called chendu-mulla meaning bouquet jasmine. Jasmine is a favorite flower of Indian ladies and is widely cultivated for flowers from which oleoresins are extracted.

NALUMANI OR FOUR O CLOCK PLANT

Scientific Name: *Mirabilis jalapa*
Common Name: Nalumani or Four O clock plant
Family : *Nyctagenacea*

This is a herb that grows on a bulb like underground stem (rhizome). It is widely grown for the beautiful bell-like flowers that appear on all seasons, the colorful part is the calyx (outer layer of petals) that is fused to form the tube. Seeds are black with wrinkled surface. This herb blooms exactly at 4 o clock gaining the name Nalumani meaning 4 O Clock in Malayalam.

NANDIARVATTOM OR CREPE JASMINE OR CARNATION OF INDIA

Scientific Name: *Tabeamaemontana diverticata*
Common Name: Nandiarvattom or Crepe jasmine or Carnation of India
Family : *Apocyanaceae*

The plant is a small shrub that produces white fragrant flowers throughout the year, leaves are simple, dark green, fruits are rarely produced; propagation done mainly through stem cuttings. These flowers are considered sacred and used for temple performances.

PALA OR BLACK-BOARD TREE OR INDIAN DEVIL TREE

Scientific Name: *Alstonia scholaris*
Common Name: Pala or Black-board tree or Indian devil tree
Family :*Apocynaceae*

A soft-stemmed tree with highly fragrant tiny flowers that open during night. Several leaves (seven, nine etc) arise from a single nod and several braches start from the same joint giving the tree a particular shape. Fruits are long pods that hang in clusters. Pala is believed to be the abode of fairies by Indian myths! The fragrance of its flower is sweet when light but causes giddiness when smelled; may be the reason for this innocent tree being called devil tree!

PITCHI, JATHIMALLI OR COMMON JASMINE OR POET'S JASMINE

Scientific Name: *Jasminum officinale*
Common Name: Pitchi, Jathimalli or Common jasmine or Poet's jasmine
Family : *Oleaceae*

A weak stemmed climber plant that bears extremely aromatic white flower clusters; this plant's leaves, roots and flowers have medicinal value. Leaves are pinnately compound with seven leaflets with a larger one at the tip. Laces made of Pitchi flowers are a craze to any Indian lady and these flowers are sold at high prices.

RAJAMALLI, DASA-MANDARAM OR DWARF POINCIANA

Scientific Name: *Caesalpinia pulcherrima*

Common Name: Rajamalli, Dasa-mandaram or Dwarf poinciana

Family :*Fabaceae*

The plant is a spiny shrub; it is a garden plant for its beautiful inflorescence with yellow and reddish flowers; leaves are compound with numerous leaflets. Pods contain five to seven seeds. As the name indicates this plant is a small shrub that does not grow more than 4 meters in height.

SANKHU-PUSHPAM, GOKARNA, APAROJITA OR BUTTERFLY PEA OR BLUE- PEA VINE

Scientific Name: *Clitoria ternatea*

Common Name: Sankhu-pushpam, Gokarna, aparojita or Butterfly pea or blue- pea vine Family : *Fabaceae*

The plant is a climber plant that bears blue conch shaped inner part above blue or white lower petal. Leaves are green pinnately compound, has much medicinal values.

The shape of conch-shell is behind the name of Shankupushpam; shanku means conch-shell in Malayalam (conch-shell is also considered sacred by Hindus).

THAMARA, KAMALAM, PADMAM OR INDIAN LOTUS

Scientific Name: *Nelumbo nucifera*

Common Name: Thamara, Kamalam, Padmam or Indian lotus

Family : *Nelumbonaceae*

It is an aquatic plant that grows in fresh water lakes and ponds with stem buried in the ground. Lotus leaves are round and floats on water connected to the stem with long stalks. Flowers are fragrant and have many red to white colored petals. Lotus is a sacred flower by Hindus and goddess Saraswati is believed to be seated on a lotus flower (Saraswati is the goddess of learning and wisdom).

THETTI, THECHI OR WEST INDIAN JASMINE OR JUNGLE GERANIUM

Scientific Name: *Ixora coccinia*

Common Name: Thetti, Thechi or West Indian jasmine or Jungle geranium

Family : *Rubiaceae*

This is an evergreen shrub that grows to less than 8 feet. Flowers are in cymose recemes; generally brick-red to yellow in color. Thetti is also a sacred plant and used in temple rituals. It is a widely grown garden plant; leaves are simple and dark green; has extensive use in Ayurvedic medicines.

THOTTAVADI OR TICKLE ME PLANT

Scientific Name: *Mimosa pudica*
Common Name: Thottavadi or Tickle me plant
Family : *Leguminosae*

It has many names such as 'touch me not, shameful plant, sleeping grass, prayer plant etc due to its property to temporarily wilt when touched (thigmotropism). It flowers profusely during the onset of summer and stays flowered until the end of autumn. Flowers are lilac with numerous stamens giving it a ball shaped beautiful shape. It is a spiny creeping plant and grows in waste lands and on road-sides. Mimosa plant has medicinal properties that can cure piles, urinary stone etc.

THULASI OR HOLY BASIL PLANT IS AN HERB WITH HIGH MEDICINAL VALUE

Scientific Name: *Ocimum tunnuiflorum* or *Ocimum sanctum*
Common Name: Thulasi or Holy basil plant is an herb with high medicinal value
Family :*Laminaceae*

It is an herb (about three feet in height) with small aromatic simple leaves; flowers terminal racemes. It is grown in front of the houses on special platforms and watered and housewives circumambulate around this plant every morning with folded hands as this plant is supposed to be a holy one.

VAKHA, GULMOHUR, KRISHNA-CHURA OR ROYAL POINCIANA

Scientific Name: *Delonix regia*
Common Name: Vakha, Gulmohur, Krishna-chura or Royal Poinciana
Family : *Fabaceae*

The plant is a weak-stemmed tree with compound leaves with numerous leaflets Its flowers are extremely beautiful with one white and other four red petals. Fruits are large sword-like pods with many seeds. A Gulmohur plant at full-bloom is an unforgettable sight.

VELIPPARUTHI OR SPANISH-FLAG IS A WILD INVASIVE SPINY SHRUB

Scientific Name: *Lantana camara*
Common Name: Velipparuthi or Spanish-flag is a wild invasive spiny shrub
Family : *Verbenaceae*

Its flowers are arranged in terminal cymes (that is born at the tip of the branch and in dumbbell shaped inflorescence). Flowers are tiny but very attractive; younger sporting lighter color.

The most important factors involved in vegetable gardening are to begin the proper plan, organize your vegetable gardening plan and then go ahead to grow vegetables.Vegetable gardening is considered as one the most favorite hobbies. If you have a small area or short space in your home, even then you can fulfill crave of vegetable gardening. Though, many people have a passion for vegetable gardening, but they are unaware of the knowledge about how to grow vegetable garden.Here are some tips to grow a vegetable garden.

VEGETABLE GARDEN

Tips To Grow A Vegetable Garden are:

1. Prepare for Vegetable gardening:To begin your vegetable gardening, you can establish a bed or place a pre-constructed frame or decorative bricks for vegetables. Before you set in the frame for vegetable gardening, the ground and soil should be leveled and weeded. It is important for vegetable gardening to prepare your soil at least three weeks before planting time to give the garden area time to mellow.
2. Consider your Garden Climate:Before choosing vegetables for vegetable gardening, it is important to know the growing season of vegetables according to the climate of your region. In south, eggplants and peppers will have a wonderful long growing season, while broccoli, peas and lettuce do well in the north. Vegetable gardening in dryer climates requires special care and maintenance to ensure that plants get enough moisture.
3. Vegetable Gardening in Small spaces:Beets are problem-free in all climates and are easy to grow. They are perfect for vegetable gardening in small area.Carrots are also considered excellent for small

space vegetable gardening. However, they don't thrive in hot weather.Lettuce is the most space saver vegetable and is perfect for small area vegetable gardening in the cooler months.Peppers not only take very little space, but will add an ornamental look to your vegetable garden.

4. Decide the location:Before knowing how to grow vegetables, it is important to decide the location for vegetable gardening. You should pick a location for vegetable gardening which is loaded with afternoon sunlight.

2

Exotic Plants in Kerala

Renowned for their showy blooms and lush foliage, our superb range of unusual and exotic plants are excellent for lending a tropical feel to your garden. Exotic-looking plants are often thought of as tender and difficult to maintain but some are very hardy in the UK! Try growing these great-value exotic flowers in borders, conservatories, or in patio containers as feature plants.

ABELMOSCHUS CRINITUS

Family: MALVACEAE

Description: Erect, perennial herb, ca 1.5 m high with fusiform tuberous taproot; young branches, petioles and pedicels simple and stellate-hairy, glabrescent later. Leaves palmately 5-7 lobed or parted, 5-8 cm across, cordate at base; lobes triangled -oblong, dentate-serrate at margin, acute or slightly acuminate at apex, hirsute on both surfaces. Petioles 1-12 cm long; stipules linear-filiform, 1-3 cm long, stellate and simple hairy. Flowers solitary or in terminal racemes by reduction of upper leaves. Epicalyx lobes 10-16, linear, 2-5 cm long, ciliate at margin, sparsely stellate-hairy; pedicels ca 2.5 cm long, accrescent.

Calyx spathaceous, 2-5 cm long, densely puberulous outside, sparsely sericeous inside. Corolla ca 6 cm across, yellow with a purple centre. Petals broadly obovoid, 4-9 x 2-4 cm, glabrous. Staminal column ca 2 cm long, glabrous, antheriferous throughout. Ovary ovoid, ca 10 x 5 mm; stylar branches ca 4 mm long; stigmas globose, purple. Capsules ovoid-globose, short-acuminate, 2-4 x 2-3 cm, hirsute. Seeds reniform or globose, ca 4 mm across with concentric rings of hairs, rarely glabrous.

Habit: Herb

Flowering & Fruiting: July-December

ABROMA AUGUSTA

Family: STERCULIACEAE

Tamil name(s): Ulatkambal

English name(s): Devil's cotton

Description: Large shrub or small tree, 2-4 m tall; branches downy. Leaves 10-20 (-30) x 5-15 (-25) cm; ovate-lanceolate, ovate-oblong, cordate at base, acute or acuminate at apex, repand-denticulate, glabrescent above, tomentose beneath; petioles 1.5-2.5 cm long; stipules linear, as long as the petiole, deciduous. Flowers few in leaf-opposed, subterminal or axillary peduncled cymes, 5 cm in diameter. Sepals ca 2 x 0.6 cm, lanceolate, connate at base, persistent. Petals ca 2.8 x 1.2 cm with hooded ca 5 mm long claw and spoon-shaped lamina, claws hairy outside. Stamens 3 in each group in the sinus; staminodes ca 2x1 mm, hairy, emarginate. Ovary 2.5-3 x 2 mm, oblong, 5- lobed, sessile; styles ca 2 mm long. Capsules 3.5-4 cm long, obpyramidal, membranous, 5-angled, truncate at apex, septicidally 5-valved, valves villous at the edge; seeds many, ca 3 x 2 mm, obovate.

Habit: Shrub to Small Tree

Flowering & Fruiting: March-August

ABUTILON STRIATUM

Family: Malvaceae

Description: Erect shrubs, 1-3 m high; young branches sparsely stellate-hairy. Leaves alternate, broadly ovate, 3-5 angled or lobed, cordate at base, serrate to subentire at margin, acuminate at apex, 6-12 x 5-11 cm, glabrous above, stellate-hairy beneath; petioles ca 10 cm long, glabrous or stellate-pubescent; stipules lanceolate, ca 3 mm long, pubescent at apex and margin at base. Flowers solitary, axillary and pendulous;.pedicels ca 12 cm long, stellate-pubescent, jointed 1 mm below flower. Calyx ca 12 mm across; lobes ovate, ca 10 x 5 mm, stellate-tomentose, glabrescent inside. Corolla red; petals obovate, ca 3 x 1.5 cm, glabrous. Staminal column ca 2.5 cm long, glabrous, antheriferous above. Ovary cylindric, ca 2 mm across, densely simple-hairy at apex, stellate-tomentose outside, 9-10 loculed; styles 10-12, glabrous; stigmas reddish, capitate.

Habit: Shrub

Flowering & Fruiting: January-April

ACACIA CATECHU

Family: Fabaceae/Leguminosae (Subfam.: Mimosoideae)
Malayalam Name(s): Karingali, Kadiram, Cutch tree
English name(s): Black catechu, Cutch tree, White kutch

Description: Deciduous, gregarious trees, to 15 m high; bark dark greyish-brown to dark brown, rough, about 1.3 cm thick, exfoliating in long, narrow rectangular strips; blaze brownish-red; branchlets brown, glabrous. Leaves bipinnate, alternate, stipulate; stipular spines slightly infra axillary, paired, 3-10 mm long, straight or hooked, occasionally lacking on flowering branchlets; rachis 8-19.5 cm, slender, pulvinate, downy, grooved above, with a gland near the base of the rachis on the upper side; pinnae 10-20 pairs, 3.6-8 cm long, opposite, downy, slender, with a gland in between the terminal 6 pairs on the lower side, leaflets 30-50, opposite, paripinnate, sessile, stipels absent; lamina 2.5-8 x 0.5-1.5 mm, linear-oblong, base unequally truncate, apex round, obtuse, or mucronate, margin entire, pubescent, chartaceous; midrib subcentral at base, lateral nerves obscure. Flowers pale yellow, sessile, in long solitary or in groups of 2-4 axillary spikes; bracts cauducous; calyx cupular-campanulate, 1-1.5 x 1.3-1.5 mm, teeth triangular or deltoid; corolla 2.5-3 mm long, lobes oblong, ovate to linear-lanceolate; stamens many, 4.5-5 mm long; ovary stipitate, 0.8-1.2 mm long, oblong-ellipsoid; style 4-5 mm long; stigma terminal. Fruit a pod 5-10 x 1-1.6 cm, flat, straight, unlobed or sinuate along margins, thin walled, beaked at apex, brown, narrowed at base into a stipe, dehiscent; seeds 3-10, orbicular or ovate, flattened.

Habit: Tree
Flowering & Fruiting: March-September

ACACIA DEALBATA

Family: Fabaceae/Leguminosae(Subfam.: Mimosoideae)
Malayalam Name(s): Wattle
English name(s): Silver Wattle

Description: Evergreen unarmed trees, to 20 m high;. bark grey, fairly smooth; branchlets and leaves grey-hoary. Leaves bipinnate, alternate, stipulate; rachis 3-9 cm, slender, pulvinate, silvery pubescent; pinnae 18-21 pairs, 0.7-2.8 cm long, slender, with a gland between each pairs on the upper side, even pinnate; leaflets 20-80, opposite, sessile, minute; lamina 2-4 x. 5-1 mm, oblong, linear or linear-oblong, base obliquely truncate, apex obtuse, margin entire, silvery pubescent, glaucous, chartaceous; lateral nerves obscure. Flowers yellow, very sweet scented, heads in axillary or terminal panicles; peduncle 5-10 mm long, ochraceous tomentellous; calyx 1-1.5 x 0.5-1.5 mm, 5-fid; corolla 2.3-3.5 mm long; petals oblong; stamens many, 3.5-4.5 mm long; ovary 4-5 mm long, subsessile, oblong, style 3-4 mm long; stigma minute, terminal. Fruit a pod 2.5-8 x 0.5-1 cm, flat, straight or curved, irregularly constricted between the seeds, acuminate at the apex, glabrous, dark reddish-brown; stalk very short; seeds 2-6, 5-6 x 3.5-4 mm, horizontal.

Habit: Tree

Flowering & Fruiting: September-March

ACACIA FARNESIANA

Family: Fabaceae/Leguminosae(Subfam.: Mimosoideae)

Description: Trees to 10 m; branchlets warty. Leaves ca. 5 in a cluster, 2-7 cm; pinnae 5 pairs, 1-2.5 cm; leaflets 10-15 pairs, elliptic, 5 x 1 mm, overlapping, glabrous, base truncate, margin entire, apex obtuse; nerves prominent below; petiole to 1.5 cm, with a gland near the middle; rachis stiff-pubescent, eglandular; stipular thorns unequal, to 2 cm, straight. Flower-heads globose, 8 mm across, 2 or 3 in axillary cluster, to 2 cm; peduncle densely stiff-pubescent, with involucel of bracteoles at the base. Flowers 2 mm across. Calyx-tube 5-toothed, to 1.5 mm. Petals 5, yellow, to 2 mm. Stamens numerous, to 4 mm, basally connate. Ovary stipitate, terete, to 1 mm; style to 3 mm. Pod terete,

5 x 0.5 cm, pulpy, turgid, glabrous, obtuse at apices, horned; seeds 20 or more, globose, 0.5 mm, 2-seriate.

Habit: Tree

Flowering & Fruiting: September-January

ACACIA MEARNSII

Family: Fabaceae/Leguminosae (Subfam.: Mimosoideae)

English name(s): Black Wattle

Description: Trees, to 15 m high; young parts silky; branchlets semiterete. Leaves bipinnate, alternate, stipulate; rachis 4-12 cm long, slender, pulvinate, pubescent, a gland at the base of the rachis on the upper side; pinnae 8-21 pairs, subopposite, 1.5-6 cm long, slender, a gland between each pairs on the upper side; leaflets 36-90, subsessile, subopposite; lamina 1.5-4 x 0.5-1 mm, linear or linear-oblong, base obtuse, subacute or obliquely truncate, apex obtuse or subacute, margin entire, puberulent, membranous, veins obscure. Flowers bisexual, white or creamy, sessile, heads arranged in axillary or terminal panicles or racemes; calyx 2.5-3 x 2-2.8 mm, ochraceous puberulous; lobes 5, ca. 2 x 1 mm, oblong; corolla 3.5-4 mm long; lobes 5, 2.5-3.5 x 1-2 mm, triangular-oblong; stamens 4-5 mm long; ovary superior, 4.5-5 mm long; style 4-5 mm long. Fruit a pod, 3-10 x 0.5-1 cm, flat, narrow, straight to slightly curved, usually constricted between the seeds, tomentose, blackish-brown.

Habit: Tree

Flowering & Fruiting: March-October

ACACIA NILOTICA SSP. INDICA

Family: Fabaceae/Leguminosae (Subfam.: Mimosoideae)

Malayalam Name(s): Karivelum, Karuvelakam, Babool

English name(s): Babul tree, Black babool, Indian gum arabic tree

Description: Trees, to 8 m high; bark black with deep narrow longitudinal fissures, running spirally up the tree, exudation gummy; blaze pinkish-brown; branchlets blackish or greyish, smooth, the young ones densely pubescent with short stiff patent hairs.

Leaves bipinnate, alternate, stipulate; stipular spines in pairs, upto 5 cm, straight, slender, white with grey spots; rachis 5-7 cm, slender, pulvinate, grooved above, pubescent; pinnae 4-9 pairs, 1.5-3 cm, slender, with a gland near the lowermost pinnae on upper side, pubescent; leaflets 20-40, opposite, sessile; lamina 3-4 x 1-2 mm, linear, oblong or linear-oblong, base oblique, apex obtuse, margin entire, glabrous, chartaceous; midrib more or less central, lateral nerves obscure. Flowers bright yellow, in heads on axillary peduncle; involucre of 2 bracteoles near or above the mid peduncle; calyx campanulate, 1-2 x 0.8-1.5 mm, glabrescent, teeth triangular to ovate; corolla 2.5-3.5 mm long, often puberulous, lobes oblong to ovate; stamens many, 4.5-5 mm long, filaments basally connate; ovary terete, stipitate, puberulous; style 5.5-6.5 mm long. Fruit a pod 7.6-23 x 1.3-2 cm, flattened, strap-shaped, straight to slightly curved, deeply constricted between seeds, bullate over seeds, beaked at apex; densely grey-felted, glaucous, greyish-green, turning black on drying, attenuate at base into stipe, dehiscent; seeds 8-13, flat 5 x 4 mm, black.

Habit: Tree

Flowering & Fruiting: July-October

ACALYPHA AMENTACEA SSP. WILKESIANA

Family: Euphorbiaceae

English name(s): Beefsteak Plant, Fire Dragon Acalypha

Description: Monoecious, compact shrub; branchlets tomentose, puberulous later. Leaves alternate, elliptic to ovate, to 25 x 12 cm, base cuneate, margin crenate-serrate, apex acuminate, glabrous except on veins, 5-nerved from base, bronzy-green-mottled with shades of red and purple; petiole to 12 cm; stipules lanceolate, to1.5 cm. Staminate spikes axillary, to 13 cm, pendulous, reddish, many-flowered. Flowers in the axils of minute bracts. Perianth 4-lobed, axillary. Stamens 8-16. Pistillate spikes erect, to 9 cm, reddish, to 12-flowered. Flowers in the axils of foliaceous, 3-angular, 9-13-parted reddish bracts. Perianth 3-5-parted; lobes ovate-lanceolate, acuminate. Ovary 3-lobed, pubescent without,

3-celled, with 1 ovule per cell, on axile placentae; styles 3, united basally, fringed apically.

Habit: Shrub

Flowering & Fruiting: October-May

ACANTHOSPERMUM HISPIDUM

Family: Asteraceae/Compositae

Malayalam Name(s): Njeringil, Musumusu

Description: Erect, densely hispid herbs. Leaves 2-4 x 1-2 cm, ovate, apex acute, entire. Heads 4-5 mm long, axillary, sessile; receptacle paleate; bracts 1-seriate, 5 mm long. Outer row of flowers female, ligulate, yellow, 3-lobed at apex. Inner flowers bisexual; corolla 2 mm long, tubular, 5-toothed at apex, yellow. Achenes 5 x 3 mm, obovate, with two diverging long spines at apex, compressed; pappus absent.

Habit: Herb

Flowering & Fruiting: January-June

ACANTHUS EBRACTEATUS

Family: Acanthaceae

Citation: Acanthus ebracteatus Vahl, Symb. Bot. 2: 75. t. 40. 1791; Hook. f., Fl. Brit. India 4: 481. 1885; Remadevi & Binoj., Journ. Econ. Tax. Bot. 24:241. 2000; Sunil & Sivadasan, Fl. Alappuzha Dist. 531. 2009.

Description: Shrubs to 1.5 m high; stem glabrous, internode 2.5 to 3 cm long, spines paired at the base of the petiole, straight. Leaves opposite, pinnatifid, c. 7.5 x 2.5 cm, serrations spinous; petiole c. 0.5 cm long. Flowers in terminal c. 3.5 cm long racemes; bract early caduceus; bracteoles 2, c. 8 x 3 mm, lanceolate, acute, puberulous along margins. Calyx lobes 4, unequal; outer 2 c. 1.5 x 1 cm, inner 2 c. 1 x 0.5 cm, pubescent along margins. Corolla tube c. 0.5 cm, lobes c. 3.6 x 3 cm, white, upper lip obsolete, lower lip elongate, obovate, shortly 3 lobed. Stamens 4, didynamous, filaments stout, curved, c. 1.5 cm and 1.3 cm long; anthers 1 celled, bearded, 1 cm long. Ovary 0.5 cm long; style 3.5 cm long, shortly bifid. Capsule ellipsoid and compressed; seeds 4.

Habit: Shrub
Flowering & Fruiting: December-April

ACANTHUS MONTANUS

Fig. More images:..>> 21,000 images, 5094 plants; 880 exotics; 100 more multiple keys for finding plants; 6,800 local/trade/common names

Family: ACANTHACEAE

Citation: Acanthus montanus T. Anderson, J. Proc. Linn. Soc. Bot. 7: 37. 1863.

ACHILLEA MILLEFOLIUM

Family: ASTERACEAE/COMPOSITAE

Citation: Achillea millefolium L., Sp. Pl. 899. 1753; Hook. f., Fl. Brit. India 3: 312. 1881; H.J. Chowdhery in Hajra et al., Fl. India 12: 5. 1995; Jain et al., Journ. Bombay Nat. Hist. Soc. 93: 458. 2000.

Description: Erect herb. Leaves alternate, linear to lanceolate, 2-3-pinnatifid, to 12 x 2.5 cm at the base. Corymbs terminal, to 10 cm wide. Capitula numerous, radiate; peduncle 4 mm. Involucre oblong. Phyllaries several-seriate, lanceolate. Receptacle convex. Outer ray florets pistillate, c. 6, rose; ligule 3-toothed. Inner disk florets bisexual. Achenes compressed. Pappus 0.

Habit: Herb

Flowering & Fruiting: August-September

ACMELLA CALVA

Family: ASTERACEAE/COMPOSITAE

Malayalam Name(s): Akravu, Eripacha, Erivalli, Kaduparni, Kuppamanjal, Naikoppu, Naimanjal, Palluvedanachedi, Tharippuchedi

English name(s): Paracress, Toothache plant

Description: Procumbent scabrid herbs. Leaves to 4 x 2.5 cm, ovate, base truncate, apex acute, chartaceous. Heads 10 x 7 mm, solitary, conical. Flowers bisexual; corolla yellow, 2.5 mm long, tubular, campanulate above; lobes 0.5 mm long, ovate, acute, glabrous. Achenes 2 x 1 mm, obovate, biconvex, dark brown, glandular.

Habit: Herb

Flowering & Fruiting: February-April

District(s): Kottayam, Idukki, Pathanamthitta, Malappuram, Palakkad, Kannur, Thiruvananthapuram, Kollam, Wayanad

Medicinal: Yes

Habitat: Moist localities in evergreen forests

Distribution: Indo-Malesia and China

POOJA FLOWERS AND DASAPUSHPAMS

TULASI (HOLY BASIL - OCIMUM TENUIFLORUM)

Tulasi is the most used pooja flower in Kerala temples. The name Tulasi means the incomparable one. A number of passages in the Puranas and other scriptures (Vedas), point to the importance of Tulasi within religious worship. Puranas like Vishnu Purana, Bhagavatha Purana, Devi Bhagavatha Purana and Padma Purana explains the glory and worshipp protocols of Tulasi in detail. Tulasi is regarded as a goddess and a consort of Lord Vishnu. Tulasi is worshipped as Tulasi Devi or Vrinda Devi. In Kerala, a household is considered incomplete if it doesn't have a Tulasithara. The month of Karthika (October-November) is famous for Tulasi pooja. Three different types of Tulasi are used in temples. They are Rama Tulasi with stems and leaves of green, Krishna Tulasi with stems and sometimes also leaves of dark green, and Karppoora Tulasi with the smell of camphor. Kattu (Vana) Tulsai (Ocimum gratissimum) the wild form, is not used for poojas. All these types are used in Ayurveda. Tulasi exhibits great variation across its range. Variations in soil type and rainfall may also equate to a difference in the size and form of the plants as well as their medicinal strength and efficacy. Puranas mentions that pooja without the use of Thulasi is incomplete and hence worthless. To know more about Tulasi click here

KOOVALAM - VILWAM, SHIVADRUMA, TRIPATRA (AEGLE MARMELOS)

Koovalam is a sacred tree to Hindus. The leaves of Koovalm are mainly used in Shiva temples. It is used while offering prayers to Lord Shiva. Koovalam leaves resembles like the three eyes of Lord Shiva, hence got the name Tripatra. Puranas like Shiva Purana, Skanda Purana, Devi Bhagavatha Purana and Padma Purana explains the glory and worshipp protocols of this tree in detail. It is planted in the premises of temples. It is related to citrus and has many names in India. Koovalam (kuvalam) or vilwmam are Malayalam names, while Bel or bael in Hindi, Kuvalum in Tamil, and Kumbala in Kannada. They are also known as golden apple, stone apple and such names in English. It is believed that Lord Shiva is present in this sacred tree. Koovalam wood is used for making Yupam in Yagams like Somayagam and Athirathram. It grows in almost all climatic conditions in wilderness. The tree is slow growing and reaches a maximum height of 8-9 meters. The pale green leaves are aromatic. The fruits can be oval or spherical and has the size of a large orange. Koovalam fruits are edible. People eat it either raw or make a good jam. Lakshmi Devi wears the fruit of Koovalm in her right hand. Hence this tree is also called Sreephalam. Koovalam fruit has astringent properties and regulate digestive functioning. It is also used in curing diarrhea and dysentery. To know more about Koovalam click here

THAMARA (LOTUS - NELUMBO NUCIFERA)

Thamara flowers are used in all temples. They are commonly used for special poojas like Sahasrakalasam, Ashtabandhakalasam and Laksharchana. Lotus has different names in Sanskrit. Some of them are Padmam, Nalinam, Aravindam, Kamalam and Pankajam.

Lotus is associated with the gods Vishnu, Brahma, and the goddesses Lakshmi and Sarasvati. From ancient times the lotus has been a divine symbol in Hindu tradition. It is often used as an example of divine beauty, for example Sri Krishna is often described as the Lotus-Eyed One. Its unfolding petals suggest the expansion of the soul. The growth of its pure beauty from the mud of its origin holds a benign spiritual promise. Particularily Brahma and Lakshmi, the divinities of potence and wealth, have the lotus symbol associated with them. In Hindu iconography, deities often are depicted with lotus flowers as their seats. Thamaramala (lotus garland) is the main offering in Koodalmanikya Swami Temple. To know more about Thamaraclick here

CHETHI, THECHI OR THETTI (IXORA COCCINEA)

Techi flower is commonly used in all temples. The flowers are found in a wide range of colours. Red, white and yellow ixora flowers are commonly used in Hindu worship. The Thechi undamala (thechi garland) of Guruvayoor Temple is famous. Only Thechi flowers are used for Poomoodal in Kadampuzha Bhagavthy Temple. It is also called as the Jungle Geranium, Flame of the Woods, and Jungle Flame.

CHEMPARATHY (HIBISCUS ROSASINENSIS)

In Sanskrit it is called Japakusumam. More than 10000 varieties of Chemparthy exists in the world. The red coloured Chemparathy flowers are usually used in Devi temples, especially in Bhadrakali temples. Light coloured flowers (white, rose, purple, yellow and orange) are used for the poojas of other deities. To know more about Chemparathy click here

NANDYAARVATTAM

Nandyarvattam is used in every temples. The flowers of Nandyarvattam are white in colour. Two different forms - idampiri and valampiri are used in temples. Nandyarvattam is used for Saraswathy pooja.

SHANKHUPUSHPAM OR APARAJITHA (BUTTERFLY PEA - CLITORIA TERNATEA)

The Malayalam word Shankhupushpam means a conch shaped flower. Butterfly pea is a deep-rooted, tall slender, climbing legume with five leaflets and a deep blue flower. Even though its origins are unknown, it is probably native to Asia according to Hortus. Shankhupushpam in two different colours - white and blue are used in temples. Blue Shankhupushpam is used for Sastha (Ayyappa)pooja. It's root is used in many Ayurvedic medicines. To know moreabout Shankhupushpam click here

THUMPA (LEUCAS ASPERA)

Thumpa flowers are very small and are white in color. They are used for Shiva pooja as well as Thrikkakkarayappa pooja on Onam day. Thumpa mala (thumpa garland) is an important offering in Kottiyoor temple.

ASHOKAM (SARACA INDICA)

Ashoka is a Sanskrit word meaning without grief or that which gives no grief. The Hindus regard it as sacred, being dedicated to Kama Deva, God of Love. One of its varieties is a very handsome, small, erect evergreen tree, with deep green foliage. Its flowers are very fragrant and are bright orange-yellow in color and later turn red. Its beautiful, delicately perfumed flowers are used in temple decoration.

MULLA (JASMINE - JASMINUM GRANDIFLORUM)

In Sanskrit it is called Mallika. Jasmine is known as the queen of flowers. The plant is known for the sweet fragrance of little milky-white flowers. Jasmine is used as religious offerings symbolizing divine hope.

They are held sacred to Vishnu and are used as votive offerings in religious ceremonies. Jasmine is a climbing vine with oval, shiny leaves and tubular, waxy-white flowers. The small white star-shaped flowers are picked at night when the aroma is most intense, so that the delicate aroma will not evaporate in the sun.

ARALI (NERIUM INDICUM OR NERIUM OLEANDER)

Arali is available in three different colours - rose, white and red. Though the flowers make great visual display, they are also highly poisonous. They are commonly used for making garlands.

CHEMPAKAM (MICHELIA CHAMPACA)

Flowers of Chempakam is usually yellow or white in colour. But there are other colours too. Only the white and yellow coloured flowers are used for poojas in Devi temples. It is not used in Shiva temples.

ERIKKU

The flower of Erikku is mainly used in Shiva temples. It is not used in Vishnu temples.

KANAKAMBARAM (FIRECRACKER FLOWER - CROSSANDRA INFUNDIBULIFORMIS)

Kanakambaram or Firecracker flower is not widely used in Kerala temples for poojas but occassionaly used for decorating the temple during festivals. It is available in orange and salmon or yellow colours. Kanakam means gold. It iswidely used for making garlands along with Jasmine.

VADAMALLI (GOMPHRENA GLOBOSA)

Vadamalli is used for decorating the temple during festivals. It is available in pink and white colors. The conelike flowerheads are beautiful in dried arrangements and will hold their shape and color indefinitely. Hence they are used for making big garlands.

RAJAMALLI (PEACOCK FLOWER - CAESALPINIA PULCHERRIMA)

Rajamalli flowers are red, orange and yellow with long red stamens. It is called Ratnagundhi colloquially. Very rarely used in temples.

CHENDUMALLI (MARIGOLD)

The common Marigold is familiar to everyone, with its pale-green leaves and golden orange flowers. The color of Chendumally range from lemon, yellow, bright yellow, golden to orange. They are very much used in making garlands for temple decoration.

JAMANTHI (CHRYSANTHEMUM - CORN MARIGOLD)

There are more than 30 types of Chrysanthemum flowers. In South India the yellow Chrysanthemum is famous as Jamanthi and are widely used for making garlands.To know more about Chrysanthemum click here

PICHAKAM OR PICHI

Pichi flowers are white in color. They are small in size with good aroma and are used in Vishnu Temples.

MANDARAM

Mandaram flowers are white in color and are used in all temples for poojas.

KAMUKIN POOKKULA

Kamukkin pookkula or flower of Aracanut tree is used for Nagadeva pooja and Kalamezhuthu. They are also used for making Palppayasam in temples like Kalampookkavu Devi temple.

DASA PUSHPAM (10 FLOWERS)

Dasa Pushpam means a group of 10 flowers. But we use leaves of most of these plants. Dasa Pushpams are used for Sheepothi Vekkal ritual in Karkkidakam, Pathirappoochoodal ritual on Thiruvathira in Dhanu, important cerimonies like marriage and for pithru karma.

KARUKA (CYNDON DACTYLON)

Karuka or Durva is widely used for Ganapathy pooja. For children with problems in Childhood in their horoscope Karuka Homa is the sure shot cure for the Balarishta yoga in the horoscope. Karuka is not used for Durga pooja.

MUKKUTTI (BIOPHYTUM SENSITIVUM)

Mukkutti or Viparitalajjalu (in Sanskrit) is a very small flowering plant. Each plant produces five to ten small flowers with yellow petals. Mukkutti is an important flower for the people of Kerala. The flower is used in athapoo, special floral formation that adores courtyards and public places during Onam, the national festival of Kerala. Mukkutti flowers and plants are used for making garlands in Ganapathy temples. Mukkutty pushpanjali is an important offering in Malliyoor Mahaganapathy temple. Mukkutty is also used for making mukkuttychanthu during Thiruvathira fesival.

NIGHT FLOWERING PLANTS FOR YOUR GARDEN

Night flowers or flowers blooming at night are a special category of flowers. They are specially for night pollinators like moths, butterflics, honeybees, etc., so while most flowers open with the first rays of the sun, these flowers open up at night.These flowers are exceedingly fragrant flowers. They attract both humans and insects with their overpowering aroma.Here are some night flowering plants for your garden.

Night Flowering Plants For your Garden are:

1. Moon Flower:This night flower is called the moon flower due to its large disc like shape and its affinity to the moon. Although it looks some what similar to its distant cousin Morning Glory, it blooms at night and is characteristically white like most of your moony garden. The plant is basically a creeper therefore fast growing and hardy.
2. Evening Primrose:t blooms at dusk and stays till dawn; these roses are white to pale pink instead of the common red rose. A very easy plant to grow and it multiplies on its own. No body needs reminding for the sweet smell of the rose that will come with it.
3. Queen Of The Night:It is basically not a plant but a variety of cactii flower that blooms once in a year and that too it might bloom once in several years. It is also known as Cereus cactus and more commonly called 'Queen of the Night' for its rare and majestic presence. It literally bleeds aroma from its beautiful white petals. If you are planting one then expect it to bloom on one special summer night.
4. Columbine:Another flower that loves the moon filled nights and opens only for the beautiful moon. White as the rest and distinct due to its blade like angular petals. Not only is the flower breathtakingly beautiful but also has a nauseatingly strong aroma. The best part is that it pollinates so well that it spreads across your garden by itself.
5. *Night Flowering Jasmine:Not all jasmines are night blooming flowers but a few are. Just like any other flowers for your night garden jasmine too is white. Jasmine plants can grow up into hardwood trees that are perennial but the night blooming variety grows in shrubs of dark green with mother of pearl white flowers in clusters. It cannot sustain in salty soil or else it is quite a adaptable plant.*

3

Area of Study, Location and Brief History of Thrissur

Man has surveyed remote galaxies and has stood on the surface of moon but has not so far come anywhere near to completing a taxonomic catalogue of the fewer than half a million species of higher plants growing on our planet (Burmmit et al., 2001). Botanists were exploring the floristic regions of the world for several centuries and their efforts have succeeded only in preparing a more realistic taxonomic account of the plants of Europe. The gravity of the situation is so severe in the tropics due to variety of reasons, the foremost being habitat destruction leading to loss of biodiversity, essential for the sustenance of life on earth. Thus, conservation of biodiversity gained prime consideration all over the world since the earth summit at Rio de Janeiro in 1992. The first and foremost process in ascertaining the biodiversity is the taxonomic treatment of living organisms.

This can be achieved only through the process involving extensive exploration, identification and documentation. Earlier works in this branch of science in the Indian subcontinent resulted in the preparation of national and provincial floras. Further studies concentrated mainly on local and regional floras. These studies resulted in the finding of several species new to science as well as species with new distribution status. This enrichment is largely due to the higher attention paid to smaller biogeographic zones. As evolution and speciation are characteristics of all living organisms and are subjected to continued variation, frequent review of the natural wealth is inevitable to update their taxonomy. It is more important for the State of Kerala because of the characteristic flora and its unique features among other biogeographic regions of the country. With the goal of preparing a comprehensive flora of India, floras of many States were completed during the last two decades. Though, the flora of most of the districts and protected areas has been studied recently, a comprehensive flora of the State has not yet been prepared. The Institute has taken up a study entitled Biodiversity Documentation for Kerala which has 12 parts and flowering plants (Spermatophyta) is one among them.

HISTORY

The History and culture of Thrissur has been developed by the influence of different cultures like Buddhism, Jainism, Brahminism and other European civilisations. Thrissur became an important centre of Sanskrit learning. The great Hindu Saint, Adi Shankara, was born very near to the city. Sankara's disciple established four Madoms (muth) in the city.

The history of Thrissur is interlinked with the Cheras of the Sangam age, who ruled over vast portions of Kerala with their capital at Vanchi. From the 9th to the 12th century Kulasekharas of Mahodayapuram took over the rule and from the 12th centuryonwards the history was of Perumpadappu Swarupam. The Perumpadappu Swarupam had its headquarters at Mahodayapuram. Central Kerala was the control of the Perumpadappu Moopil, known as the 'Kerala Chakravarthi'. In the 14th and 15th centuries the Zamorins of Calicut managed to occupy a large part of the present Thrissur district. Kodungalloor, the old harbour of India, attracted European powers to Kerala for trading spices and other commodities. In the consequent centuries the European powers dominated the scene. In the start of 15th century, the Portuguese came and by the beginning of the 17th century the Dutch and the English appeared on the panorama and challenged the Portuguese. Internal conflict in the Perumpadappu Swarupam helped the Dutch to establish their domination in Kerala coast. Hyder Ali and Tippu Sultan figured very prominent in the northern part of Kerala during that period. The history of modern Thrissur started with the swearing ceremony of Raja Rama Varma in 1790, who is popularly known as Sakthan Thampuran (1790-1805). The mighty ruler Sakthan Thampuran is legendary as the architect of Thrissur town.

From ancient times, Thrissur district has played a significant part in the political history of south India. The early political history of the district is interlinked with that of the Cheras of the Sangam age, who ruled over vast portions of Kerala with their capital at Vanchi. The whole of the presentThrissur district was included in the early Chera empire.

The district can claim to have played a significant part in fostering the trade relations between Kerala and the outside world in the ancient and medieval period. It can also claim to have played an important part in fostering cultural relations and in laying the foundation of a cosmopolitan and composite culture in this part of the country. Kodungalloor which had the unique distinction of being the 'Premium Emporium India', also belongs to the signal honour or having first given shelter to all the three communities which have contributed to the prosperity of Malabar'. These three communities are the Christians, the Jews and the Muslims.

The history of Thrissur district from the 9th to the 12th centuries is the history of Kulasekharas of Mahodayapuram and the history since 12th century is the history of the rise and growth of Perumpadappu Swarupam. In the course of its long and chequered history, the Perumpadappu Swarupam had its capital at different places. According to the literary works of the period, the Perumpadappu Swarupam had its headquarters at Mahodayapuyram and had a number of Naduvazhies in southern Kerala. Central Kerala recognised the supremacy of the Perumpadappu Moopil and he is even referred to as the 'Kerala Chakravarthi' in the 'Sivavilasam' and some other works.

One of the landmarks in the history of the Perumpadappu Swarupam is the foundation of a new era called *Pudu Vaipu* era. The *Pudu Vaipu* era is traditionally believed to have commenced from the date on which the island of Vypeen was thrown from the sea.

The 14th and 15th centuries constituted a period of aggressive wars in the course of which the Zamorins of Calicut acquired a large part of the present Thrissur district. In the subsequent centuries the Portuguese dominated the scene. By the beginning of the 17th century the Portuguese power in Kerala was on the verge of collapse. About this time other European powers like the Dutch and the English appeared on the scene and challenged the Portuguese. Internal dissension in the Perumpadappu Swarupam helped the Dutch in getting a footing on the Kerala coast. As the Kerala chiefs were conscious of the impending doom of the Portuguese, they looked upon the Dutch as the rising power and extended a hearty welcome to them.

The decadence and consequential want of solidarity opened the flood gates of aggression. Hyder Ali and Tippu Sultan figured very prominently during the period. In 1790 Raja Rama Varma (1790-1805) popularly known as Sakthan Thampuran ascended the throne of Cochin. With the accession of this ruler the modern period in the history of Cochin and of the district begin. Sakthan Thampuran was the most powerful*maharaja* as the very name indicate. He is the architect of Thrissur town. Sakthan Thampuran ascended the throne just before the conclusion of a treaty with the English East Company. According to that treaty, Cochin threw off all allegiance to Tippu and became a tributary to the Company. The wave of nationalism and political consciousness which swept through the country since the early decades of this century had its repercussion in the district as well.

Even as early as 1919 a committee of the Indian National Congress was functioning in Thrissur. In the Civil Disobedience Movement of 1921, several persons in Thrissur town and other places in the district took active part and courted arrest. Thrissur district can claim the honour of having been in the forefront of the countrywide movement for temple entry and abolition of untouchability. The famous *Guruvayur Satyagraha*is a memorable episode in the history of the National Movement.

The Government of Cochin under the guidance of Sri. R. K. Shanmughom Chetti followed a policy of conciliation. By decree the public demand for the introduction of responsible Government in the State grew strong. In August

1938 Cochin announced a scheme for reforming the State legislature and introducing a system as per the Government of India Act of 1919 in the British Indian provinces. The administration of certain departments was entrusted to an elected member of the legislature to be nominated by the Maharaja. In the elections to the reformed legislature two political parties, viz. the Cochin State Congress and the Cochin Congress won 12 and 13 seats respectively. With the help of a few independents Ambat Sivarama Menon who was the leader of the Cochin Congress Party took up office as Minister under the scheme in June 1938. On his death in August 1938 Dr. A.R. Menon was appointed as Minister. When the State Legislature passed a vote of non-confidence against him, Dr. Menon resigned office on February 25,1942 and was succeeded by Sri T.K. Nair who continued in office till July 11,1945.

The introduction of dyarchy did not satisfy the political aspirations of the people of Cochin. The idea of full responsible Government on the basis of adult franchise had caught their imagination. On January 26, 1941 a new political organisation called the Cochin State Praja Mandal took shape on the initiative of a few young politicians under the leadership of V.R. Krishnan Ezhuthachan.

The 'Quit India' Movement of 1942 had its echoes in the district. After the release of the leaders from jail in 1943, the Cochin State Praja Mandal pursued its organisational activities more vigorously. In the elections to the State Legislature in 1945 it won 12, of the 19 seats contested by its candidates. At the annual conference of the Praja Mandal held at Ernakulam in 1946 it was decided to start a state wide movement for the achievement of a responsible Government. The State Legislature was scheduled to meet on July 29, and it was decided that the day should be observed all over the State as 'Responsible Government Day'. In pursuance of this decision, meetings and demonstrations were held all over the State demanding the end of Dewan's rule and the transfer of full political power to the elected representatives of the people. The Maharaja of Cochin announced in August 1946 his decision to transfer all departments of the State Government except law and order and finance to the control of Ministers responsible to the State Legislature. In co-operation with other parties in the State Legislature, the Cochin State Praja Mandal decided to accept the offer. Consequently the first popular Cabinet of Cochin consisting of Panampilli Govinda Menon, C.R. Iyyunni, K. Ayyappan and T.K. Nair assumed office.

The first step towards the achievement of the goal of '*Aikyakerala*' was taken with the integration of 'Travancore Cochin' States in July 1949. With the linguistic reorganisation of States in India, in November 1956 the Kerala State came into existence

GEOGRAPHY

Thrissur is situated in south western India, 10.52° N and 76.21° E and lies in the central part of Kerala, India. The area is about 66.15 km2. The centre of the city is the Swaraj Round, where the famous Vadakkumnathan temple and Thekin Kaadu Maidan (Teak forest) are situated. It is also the venue of the famed colourful cultural event, the Thrissur Pooram. The city enjoys tropical climate; monsoons start in June. The months of April and May can get pretty humid. The best weather is from October to February.

The city lies at 10.52°N 76.21°E and has an average altitude of 2.83 metres. The city is located in midland regions of Kerala, with an extended part of Palakkad plains. It is situated in hillock, which allows rain water to automatically drain out of the city. The city geologically is composed of Archaean gneisses and crystalline schists. Major parts of city are covered by Archaean rocks. Thrissur lies near the center of the Indian tectonic plate (the Indian Plate) and is subject to comparatively little seismic or volcanic activity. The predominant topography of the city is Thrissur Kole Wetlands which extends to Malappuram district. The city's highest point is Vilangan Hills. The city is dotted with ponds and small rivers which act as a natural drainage system for the city. Peechi-Vazhani Wildlife Sanctuary inWestern Ghats near Thrissur city act as a source of water for the city.

Climate

Under the Köppen climate classification, the city features a Tropical monsoon climate. Since the region lies in the south western coastal state of Kerala, the climate is tropical, with only minor differences in temperatures between day and night, as well as over the year. Summer lasts from March to May, and is followed by the South-west monsoon from June to September. October and November form the post monsoon or retreating monsoon season. Winter from December through February is slightly cooler, and windy, due to winds from the Western Ghats. The City is drained in the monsoonal season by heavy showers. The average annual rainfall is 2500 mm. TheSouth-west monsoon generally sets in during the last week of May. After July the rainfall decreases. On an average, there are 124 rainy days in a year. The maximum average temperature of the city in the summer season is 33 degree Celsius while the minimum temperature recorded is 22.5 degrees Celsius. The winter season records a maximum average of 29 degree Celsius and a minimum average of 20 degree Celsius.

CULTURE

Thrissur, the cradle of culture of Kerala is decorated with Vadakkunnathan (Siva) Temple, from where the name derived. The old name was 'Thrishivaperur' means the place with 'Lord Shiva's name'. The Thrissur Pooram, celebrated during April-May is top on the list of cultural attraction of the city. The Kudamattom (Umbrella Display competition), drum concerts and the spectacular fireworks all add colour to the Pooram celebration.

The significant culture promoting institutes like Kerala Sahitya Academy, Kerala Sangeetha Nataka Academy and Kerala Lalithakala Academy are head quartered at Thrissur to promote the growth of Malayalam language, literature, dance, music, drama, folk arts, fine arts and artistic heritage of Kerala. The famous Kerala Kalamandalam is also situated in Thrissur District.

Aryabhata, the world famous mathematician and astronomist was believed to be born in Kodungallor. Thrissur give birth to famous thinkers and writers such as Joseph Mundassery, C. Achuthamenon, P. Bhaskaran and many other progressive philosophists. St. Thomas, the renowned Christian missionary and Malik Ibn Deenar, the messenger of Allah were visited Thrissur, who have established Christianity and Islamism respectively in India. Sree Narayana Guru, Swami Vivekananda and Mahatma Gandhi have visited Thrissur, which helped people to fight against social injustices in the caste-ridden society of Kerala.

Thrissur's contribution to the field of Science and Technology is also outstanding. The famous Research Institutes like Kerala Forest Research Institute (KFRI, Peechi), Kerala Agriculture University (KAU, Mannuthy) and Amala Cancer Research Institute are standing mainstay in this stream.

ECONOMY OF THRISSUR

The city of Thrissur, the *Cultural Capital of Kerala*, is also considered as a major commercial and business hub of South India. It is said to be the heartland of Kerala's business acumen and home to most every leading *Malayali* entrepreneurs. The city which is famous for *Bullion, Banking and its Business* acumen, is the darling of investors in Kerala. Thrissur city is also referred as the *Golden city of India*. It manufactures 70% of plain gold jewellery in Kerala per day. According to a survey, Thrissur city has been placed on 7th among the ten cities in India to reside. According to Registrar of Companies, the period from January 1 to March 31, 2010, 87 companies were registered in Thrissur city and stood second in Keralaafter Cochin. Thrissur's traditional strength lies in best entrepreneurial and financial capabilities.

HISTORY

Historians say that King of Cochin, Sakthan Thampuran, invited 52 Syrian Christian families from the neighboring Pandya Kingdom, established them at *oottupura* or mass feeding centers and encouraged them to do business in textiles. These businessmen, with their traditional flair for trade soon built up City of Thrissur into the most flourishing centre for internal trade in South India. Their financial acumen has been mainly responsible for founding and building up the Kuri system of financing which has now become an all India institution. This has also made Thrissur city the most important banking centre in South India. The industries of Thrissur city contribute largely to the district's and city's economy. The presence of leading industrial houses in Thrissur city has boosted the industrial scenario of the area. Though, the city has numerous types of industries inThrissur. The economy of Thrissur is largely dependent on industries, retailing and financing. In fact, Thrissur is one of the most important industrial centers of the state of Kerala. Industries like textile, timber, coir, fishery industries, agriculture-based industries, tiles industries are present in Thrissur. However the above mentioned industries are not the only sources of revenue for Thrissur. The tourism industry plays an important role in Thrissur's economy.

Retailing

As Kerala is known more as a consumer state rather than a producer state,Thrissur also carries the same tag, as retailing is a big business and revenue earner for the City. Jewellery and textile retailing occupies a major part of the retailing business in Thrissur. The City is considered as hub of jewellery and textile business in South India. Most of the jewellery groups have outlets in Thrissur and provide jobs to thousands of people. Kalyan Group, Kalyan Silks, Kalyan Jewellers, Jos Alukka & Sons, Josco Group, Seemas Wedding Collections, Pulimoottil Silks, Emmanuval Silks, Sree Lakshmi Silks, Fashion Fabrics, Elite Fabrics, Elite Sareee House, Modern Silks, Manshire, Lakshmi Silks, Kalima Collections and Chakola Silkhouse are the few to name.

Tile industry

One of the significant industries of the area, the tile industry of Thrissur employs numerous labors from the state and outside. TheThrissur tile industry is more than a hundred years old. Over the years the tile industry in Thrissur has witnessed many rise and fall but it has managed to withstand the bad times. Today, Thrissur can boasts of 160 tile factories making it one of the biggest hubs of southern India. The tile industry of Thrissur is aided by the presence of clay suitable for making tiles. The main centers of Tile Industry in Thrissur are in Karuvannur, Pudukad, Ollur and Amballur. The tile industry has acquired Italian technology to improve the quality and production of tiles. The tile industry provides many jobs to the local people of the city.

Banking and Finance

The main strength of Thrissur's economy is its financial capabilities. The banking system in Thrissur have been accredited for the effective disbursement of credit to the various sectors of the Kerala society. It had played a big role in Kerala's economy from the earlier part of 1800's. It was started by Christian traders in the form of Chit fund. Gradually, Chit fund contributed much to the development of the banking industry in Thrissur. According to Reserve Bank of India, Thrissur in the 1930s boasted of head offices of 58 banks and was recognised by RBI as 'Banking town of India. Now, Thrissur boast the headquarters of four scheduled banks in Kerala, South Indian Bank Ltd, Catholic Syrian Bank, Dhanalakshmi Bank and Lord Krishna Bank. The city also is the headquarters of Manappuram General Finance and Leasing Ltd, Kerala State Financial Enterprise and ESAF Microfinance and Investments, leading non-banking finance companies in India. According to Reserve Bank of India, Thrissur in the 1930s boasted of head offices of 58 banks and was recognised by RBI as *Banking town*. According to All Kerala Kuri Foreman's Association, Kerala have around 5,000 chit companies, with Thrissur accounting for the maximum of 3,000. These chit companies provide employment to about 35,000 persons directly and an equal number indirectly.

Ayurvedic Drug Manufacturing Industry

Thrissur is also emerging as the largest hub for Ayurvedic drug manufacturing industry in theIndia. Out of the 850 ayurvedic drug-manufacturing units in Kerala, about 150 units, including some of the major ayurvedic drug manufacturers in the state are located in and around theThrissur. Of these, some of the companies like the Oushadhi, Vaidyaratnam Oushadhasala,KP Namboodiris, Sitaram Ayurvedic Pharmacy Ltd, Kandamkulathy Vaidyasala, SNA Oushadhasala etc. are among the leading ayurvedic drug manufacturers in the state.

Thrissur Ayurveda Cluster, anonther initiative by a group of Ayurvedic manufacturers of Thrissur, has developed a cluster in KINFRA Park in Koratty in Thrissur District. The cluster is meant for a comprehensive development of Kerala brand of Ayurvedic products and train the manufacturers of Ayurveda products on the importance of safety, quality and efficacy. The cluster have all the state-of the-art facilities for testing and analysis, process product validation, safety study and manufacture. The cluster is approved by the Department of Ayurveda, Yoga & Naturopathy, Unani, Siddha and Homoeopathy (AYUSH).

Textile Industry

The prosperous textile mills have made the textile industry of Thrissur a leading industry in South India. At present there are six such mills in Thrissur. The textile mills in Thrissurare Sitaram Spinning and Weaving Mills; Thrissur, Alagappa Textiles, Alagappa Nagar; Kerala Lakshmi Mills, Pullazhi; Cotton Mills, Nattika; Rajgopal Textiles, Athani; Kunnath Textiles and Vanaja Textiles in Kuriachira. Mainly hosiery products are manufactured in most of these mills. Thrissur is one of the leading producer of hosiery end products in theKerala. Popular as sewing threads, Vaiga is sold in every corner of the country. The first mill in Thrissur was the Sitaram Spinning and Weaving Mills. It was established in 1909 but was devastated by fire in 1953. However, it was soon reorganized and today it is one of the main textile plants of the area.

Gold

Thrissur city can also be referred as the *Gold Capital of India*, since there is around Rs 700 crore business of gold every year in the city. All major jewelleries in Kerala have branches in city. It is one of the main manufacturing centers of plain gold jewellery in theSouth India. 70% of Kerala's jewellery is manufactured in this city. Gold jewellery in Thrissur is well known for its world class craftsmanship and exquisite dexterity craftsmanship.

Diamond Industry

Thrissur is known as the hub of diamond and gold jewellery production in South India. Majority of the units located in Kapiparambu, Tholur, Adat, Choondal and Avanur in Thrissur Metropolitan Area. There are around 75 units employing around 5,000 people. The diamond polishing work was brought to Kerala by P K Sankunni in 1964 from Mumbai. It flourished in these areas providing employment to many in their own villages especially to women. There is flexibility in the working hours as the workers are living near the units. Because of higher wages and less encouragement from Kerala Government, most of the traders are relocating to Surat inGujarat. At the height of the business there were 310 units and around 25,000 workers were involved in this business.

Thrissur Pooram

According to estimates, around a million people around the world witness the annual Thrissur Pooram. The two main participants of the 36-hour festival, Thiruvambadi and Paramekkavu temples, spend about Rs 1.2 crore on the festival. Hotels and flats in the city give rooms and terraces for people to witness the festival on rent, charging Rs 750-1,000 per person. Liquor sales, too, see an increase during the festival. Sales through the various liquor shops and bars in the Thrissur city were worth around Rs 72 lakh on the Pooram day alone in 2008, according to officials of Kerala State Beverages Corporation, the Government-owned liquor retailer in the state.

Real Estate

Real estate is one of the sunshine sectors of city of Thrissur's economy. Land availability, image of a clean and green city, good law and order situation, presence of reputed educational institutions and tremendous scope for promoting tourism industry are the pillars of Thrissur's real estate hopes. With Kochi becoming more and more congested day-by-day, in due course of time the city of Thrissur is expected to see a spurt in the number of property buyers. City of Thrissur has always been a favourite among builders of repute as it is in the centre of Kerala state.

Further, the tourist traffic to the adjoining Guruvayur Temple is demanding an increase in accommodation facilities.

Leading property developers like Southern Investments, Thrissur Builders, Cheloor Group, Sreeramajayam Builders, Jos Alukkas Group & Developers, Maya Realtors, Skyline Builders and Sobha Developers Ltd are the major players. According to real estate consultants Cushman & Wakefield, the presence of a number of educational institutions, leading hospitals are instrumental in driving the demand for residential property. A Cushman report says that the availability of land and preference to villas have convinced developers to go in for niche products, even though apartments are coming up in key locations where land is scarce.

Kuries

Thrissur is also known for its Kuries or Chit companies. Around 2,000 such companies are located in Thrissur city itself both government and private. These Kuri companies have played a major role in developing Thrissur city into economic hub in South India.

FESTIVALS

Thrissur is the land of pooram. Thrissur Pooram, the pooram of all poorams, is in the top of that list. Arattupuzha Pooram, Machadu Mamangam, Kodungallur Bharani are some of the other famous festivals in Thrissur.

THRISSUR POORAM

Thrissur/ Thrissivaperoor pooram commonly called 'the pooram of all poorams'. It is one of the mammoth pooram festival in kerala. It has been celebrated every year in Medam (April) month, as per malayalam calendar, at Vadakkumnathan temple, situated on famous *thekkinkadu maidhanam*, a hillock right in the centre of city. It was believed that every year the dynastic gods and goddesses of neighbouring temple met together for a day of celebration. It is a 36 hours continuous pooram attracting a huge mass of people from different places including international tourists. The two dewasoms – Thiruvampadi and Paramekkavu are the two temples are the major temples to make the festival a remarkable one.

The pooram starts at the time of Kanimangalam sasthavu ezhunnellippu in the early morning and followed by the ezhunnellippu of other six temples. One of the major event in Thrissur pooram is "*Madathil varavu*"- is a panchavadhyam melam, participatingmore than 200 artists, consists of *Thimila, Madhalam, Trumpet, Cymbal and Edakka* (Different types of instruments). At 2' O clock, inside the vadakkumnathan temple starts the famous *Ilanjithara melam* – a type of melam consists of drum, trumpets, pipe and cymbal.

Just two days before pooram, there is a huge exhibition – *Ana chamaya pradharshanam* (exhibition of elephant decorations), of both temple at various schools.

The pooram has a good collection of elephants (more than 50) decorated with *nettipattam* (decorative golden headdress), strikingly crafted *Kolam*, decorative bells, ornaments and the umbrellas, venchamaram, and alavattam are awesome and it enrich the beauty of elephants and pooram.

At the end of the pooram, after the Ilanjithara melam, both Paramekkavu and Thiruvambadi groups enter the temple through the western gate and come out through the southern gate and array themselves, face to face in distant places. The two groups in the presence of melam, exchange colourful and crafted umbrellas competitively at the top of the elephants – called *Kudamattom*, which is eye catching attraction of the pooram.

The pooram concluded with a spectacular fire works display, which is held in next day early morning after the pooram. The two temples competitively crack many innovative and charming fire works, which make spectators going into ruptures.

The another notable feature of the pooram is its secular nature. All other communities actively participate and make their prominent role in each and every part of the festival. Most of the pandal works are crafted by muslim community. The materials for the umbrellas for '*Kudamattom*' are offered by the churches and their members. It is a good sign of secularism which is disintegrating nowadays.

This pooram festival differs from other national festivals like Kumbha Mela of Uttar Pradesh, the Vijayadashami pageantry of Mysor or the Rath Yatra of Orissa. This pooram upholding communal harmony and all people from different religion gives hand to hand to the success of pooram.

HISTORY OF POORAM

Thrissur pooram is started two centuries back the then ruler of Cochin, Sakthan Thampuran or Raja Rama Varma, in 1798. Sakthan Thampuran, so known for his firm and decisive administration, decided to break tradition and started to celebrate the pooram festival belonging to his region. Before the initiation of Thrissur pooram, Arattupuzha festival was the largest temple festival, which is around 12 Km from the city. Temples near the Thrissur were the regular participants of the Arattupuzha pooram untile they were denied by the chief of Peruvanam Gramam due to the delayed entry of the Thrissur and Kuttanellur termple. This caused the Thrissur Naduvazhi, the chief poojari of Vadakkunnathan, known as Yogadiripad and the Kuttanellur Naduvazhi started the pooram in Thrissur.This pooram started as an act of reprisal quickly lost its charm, after infighting between the two main Naduvazhis. It required the intervention of the ruler to get this right. Sakthan Thampuran unified the 10 temples situated around Vadakkunnathan temple and organized the celebration of Thrissur Pooram as a mass festival.

He ordained these temples into two groups, Western group and Eastern group. The Western group as Thiruvambady consisting of Kanimangalam, Laloor, Ayyanthole, Nethilakkavu and the Thiruvambady temple, as the main one. The Eastern group called as Paramekkavu, consisting in addition to Paramekkavu temple, Karamukku, Chembukavu, Choorakottukavu and Panamukkamppilly. The pooram was to be centered around the Vadakkunnathan temple, with all these temples sending their poorams (the whole procession), to pay obeisance to the Shiva, the presiding deity. The Thampuran is believed to have chalked out the program and the main events of the Thrissur pooram festival. It is this historical background that determines the course of the pooram program and it is specifically the ruler's antipathy to the Brahmin aristocracy to open Thrissur pooram for the common man.

PEOPLE

The total population of Thrissur district according to the census of 2001 is 2,975,440 of whom 1,422,047 are men and 1,553,393 are women. Hindus constitute the bulk of the population of this district. Other communities are Christians and Muslims. The Konkani Brahmins are another immigrant caste and they are found mainly in Cranganore and Mukundapuram taluks.

The Nairs who till recently followed the *Marumakkathayam* family system constitute the most important section among the Hindus of Thrissur. Now a vast majority of them have taken to agriculture while others have been absorbed in Government service and other professions. Till a few decades ago, the Nairs were divided into several sub-castes and inter-dining and inter-marriages were not permitted among them. The Nairs attached to Namboodiri and Kshatriya houses for certain domestic and religious services were called Illathu Nairs and Swarupattil Nairs respectively. Charna Nairs, Pallichans, Vattekadans, Odathu Nairs, Auduru Nairs and Attikurussi Nairs are other Nair subdivisions. Every Nair had a title affixed to his name. Achan, Kartha, Kaimal and Mannadiar were some of the titles of nobility conferred on the Nairs by the Rajas of Cochin while Panikkar and Kurup were the titles of those who maintained Kalaries as their hereditary profession. Menon was the title conferred on the Nairs who followed a literacy career. When the country underwent tremendous changes, strict observations of caste rules fell into disuse.

The Samanthans, though very few in number in the district, are said to have sprung from the union of Kshatriya men with Nair women. They have *marumakkathayis*. The Ezhavas who follow Makkathayam are numerically one of the strongest communities in Thrissur. They have attained important positions as merchants, landowners and cultivators. A good number of them have also taken to learned professions. Velythedans, Velakkattalavans and Chaliyans are hereditary washermen, barbers and weavers respectively. Ezhuthachans otherwise known as Kadupottans who follow the patriarchal system of inheritance are supposed to be the descendants of Pattar Brahmins. They are hereditary village school masters. The Valans, Arayans and Mukkuvas are fishermen mostly living in the coastal areas of Thrissur district. Besides, there are a number of other castes like the Mannans, Velans, Pulluvans and Pattilans in the district.

Another section among the Hindus is the Kammalas who are divided into carpenters, masons, braziers, blacksmiths, goldsmiths etc. As their service is essential, they are till engaged in their traditional occupations. But in recent years a sizable section of them have taken to modern education and steady progress. The Devanga Chettis and Kaikolans are weaving castes found in Mukundapuram taluk. They immigrated into the district from Mysore and Coimbatore respectively. The Vaniyans Kudumis, Pandithans, Kallans, Pandarams, Ambattans, Vannans, Chakkiliyans and Kusavan are also immigrant castes. The Vaniyans wear the sacred thread and resemble Konkani Brahmins. The Pandarams are engaged in making Pappadam, the favourite crisp cake of the Malayalees. Ambattans are Tamil barbers and Vannans are Tamil washermen.

The Scheduled Castes and Scheduled Tribes form a significant section among the Hindus of the district. The former are mainly agricultural labourers and are found in all the Taluks. The Scheduled Tribes of the district are the Kadar, the Malayar and the Mathuvans. The Kadar of the area belong to two clans, the Anamala Kadar living at Parambikulam and West Kadar living at Adirappilli. The Malayar and Kadar are nomadic people.

Christians form the second largest community in the district. It is strange that Cranganore where the Gospel of Christ is believed to have been first preached in India should have the lowest proportion of Christians among the taluks of the district. The earliest Church in the district was a Nestorian branch of the Asiatic Church presided over by Bishops usually ordained in Persia. The early Christians were known as Syrian Christians. The Syrians Catholics, Latin Catholics, Jacobites, the Reformed Syrians and Protestants are some of the main sections of the Christian Community in the district . In Thrissur and its neighbourhood there is a small community of Christians known as Chaldeans.

The Christians have a predominant place in the social and economic life of the district. Trade and agriculture are the chief occupations of the community. There have been several survivals of Hindu customs among the Christians such as caste prejudice, belief in astrology, omens, witchcraft and charms, the tying of the tali as part of the marriage ceremony and its removal on the death of the husband, the performance of Sradha or the annual ceremony for the soul of the dead etc.

Muslims form the third major community in the district. A majority of them are found in Chavakkad and Kodungalloor taluks. Most of them are Sunnis. Some of the Muslims are cultivators or traders, while the majority are boatmen, fishermen and labourers of every description.

CUSTOMS AND RITUALS

Serpent (naga) worship and ancestor worship, evidently non-Aryan practices, have been widely prevalent in the district. The temples here are centres of religious activity. The Vadakkunnathan Temple at Thrissur, Koodalmanikam temple at Irinjalakuda, the Kurumba Bhagavathi temple at

Kodungalloor, the Sri Rama Temple at Triprayar, the Sri Krishna Temple at Guruvayoor are some of the reputed shrines. The prominent Gods and Goddesses worshipped are *Vishnu, Siva, Bhagavathi, Siva, Bhagavathi, Subramonia* and*Sastha*.

Fasting is a significant form of religious observance. It is observed on *Shashti, Ekadasi, Pradosham*, Full Moon and New Moon days. On Shashti viz. the sixth day of the fortnight, fast is observed by those who wish for issue. Ekadasi is sacred to Vishnu and Pradosham to Siva. Fast in honour of the Goddess Parvathi is observed on Full Moon days. The observance of festivals is an important aspect of religious activities. Here special mention may be made of the three major festivals of the Malayalees viz. Onam, Vishu and Tiruvathira. Among the ceremonies still current may be mentioned N*amakaranam, Chorunu, Vidyarambham, Upanayanam* and *Sradha*.

The laws of inheritance prevalent in the district have been the *makkathayam* (Patrilineal) and*Marumakkathayam* (Matrilineal system). *Marumakkathayam* is the dominant one which most of the people were in allegiance. The *Ambalavasis, the Kshatriyas, Samantans, Velekkattavans, Veluthedans* and a few other castes have followed the *Marumakkathayam* system in the district. Among the communities that have followed the *Makkathayam* system may be mentioned the *Namboothiris, Ezhavas, Kammalas, Kanakkans, Cherumans, Tanda Pulayans, Vettuvas, Ezhuthachans, Kanisans, Panans, Perumannans, Mannans, Velas, Velans, Arayans, Amukuvans, Mukkuvans, Marakkans* and all the hill tribes. The Christians and Muslims also have been *Makkathayis*.

THRISSUR POORAM

The famous Thrissur Pooram is an annual festival celebrated during April-May in the Vadakkumnathan temple here. During the festival idols of Gods and Goddesses from various temples are brought in all pomp and pagentry with the play of drums and musical instruments and pro-technics to the Thekkinkadu Maaidan. Lakhs of people attend the festival every year. An all India Exhibition is also conducted every year during the Pooram days under the combined auspices of the Thiruvampady and Paramekkavu Devaswoms at the temple premises.

AGRICULTURE

Rice , tapioca, coconut, areacanut, rubber, cashew and banana are the most important agricultural products of the district. The most important crop of the district is paddy. In certain areas three crops are raised (Viruppu, Mundakan and Punja) in a year. One of the striking features in regard to agricultural operations in the district is the cultivation. Extensive low-level lakes in Thrissur and Mukundpuram taluks are artificially reclaimed and bunded. Tapioca is the food of the poor and the middle class. The reason for the large scale consumption of tapioca is attribute to its high calorific value. Coconut is one of the important garden crops of the district. Among condiments and spices the arecanut tree stands first. Poor farmers with small holdings are the cultivators of the crop.

Fruits and Vegetables

Thrissur is a land of fruits. Perhaps no other district in the State grows a greater variety of fruits or has better facilities for horticulture. Plantain, jack fruits, mangoes, bread fruits, pineapples, etc., are grown in abundance in most parts of the district. Jack and mango trees are extensively grown in the gardens attached to houses. Cashewnut is cultivated in almost all parts of the district. The cultivation of rubber is popular.

Animal Husbandry

The district affords the best example to the fact that a damp climate is not conductive to the growth of cattle. The indigenous breed of cattle is weak and stunted in growth.

CULTURAL TRADITION

The cultural tradition of the district goes back to very early days. There were great centers of learning and culture in the district in the ancient and early medieval periods. In the early centuries of the Christian era, Mathilakam was a great centre of learning and culture. Buddhist and Jai scholars of repute are said to have lived here and engaged themselves and teaching. At a later district in the ancient and early medieval periods. In the early days. There were great centres of learning and culture in the stage, under the Kullasekhara of the second Chera Empire, Mahodayapuram became famous as a great seat of learning and culture. The greatest literacy figure in the district was Mahakavi Vallathol Narayana Menon. Though born in the Malabar region of Kerala. Vallathol made Cheruthuruthy his headquarters. He was not only a great poet but also a distinguished patron of the arts of Kerala, particularly Kathakali. He founded the Kerala Kalamandalam of Cheruthuruthy to disseminate the art and culture of Kerala.

FISHERIES

Thrissur district has a long tradition in the field of fishing industry. If offers natural facilities for marine and inland fisheries. Its coast line is about 54 km. in length from Azhikode to Puthenkadappuram. Fishing is the main occupation of a large number of people. The main fishing castes are Valan, Aryan, Mukkuvan and Marakkan. Thrissur is one of the biggest fish market of Kerala. Fish in an important item in the diet of about 90% of the population. Oil sardines are used as manure. About 95% of the total catch is marketed within the district. The fishing industry thus makes a sizable contribution to the wealth of the district, and is the main source of income of a large section of the people inhabiting the coastal area. There are seven major fishing centres in the district viz., Azhikode, Nattika, Vadanappilli, Kadappuram, Blangad, Puthenkadappuram and Chettuva. The district have 18 coastal fisheries villages and three inland fisheries villages. There is a Shrimp hatchery at Azhikode.

FORESTS

A total area under forests in the district is 1006.72 sq. km. The forests of the district are mainly seen in the eastern portion of Talappilli, Thrissur and Mukundapuram taluks. They extend from the Shoroor river (Bharathapuzha) in the north to the Chalakudy river in the south. The Initial works of a wild life sanctuary has been started at Echhippara in the reservoir area of Chimmani dam. A tree park with facilities to conduct studies on trees and forest for public functioning at Kuthiran under social forestry is a unique instance in the state. This institution is the second one in the country.

Forest produce: The chief forest produce is timber. The principal local markets for timber are Cochin, Ernakulam and Thrissur. A large quantity of timber is transported to Coimbatore and Pollachi. High girth rosewood is exported to foreign countries. Other hard wood species which command a steady market are Irul, Pullamaruthu, Koramaruthu, Venga, Venteak, Pongu, Agil etc. Minor forest products are also abundant in the district. Mattaipal Karuvelampatta, Marotti, Poovam, Zamalporia, Kanjiram, Elavarngam are some of them. Kerala Forest Research Institute (KFRI) at Peechi is an institution which conducts ecological and Forests Development Research studies. The Institute has a very good nursery of medical plants.

INDUSTRIES

Power loom Industry: There are six power loom factories in Co-operative sector in the district. They are at Kodungallur, Aviniseri, Adat, Machad, Nadathara and Manaloor. In addition to this, there is an institutional Power loom complex at Keecheri.

Textile Industry: There are six textile mills in the district. They are Alagappa Textiles at Alagappanagr, Kerala Lakshmi Mills at Pullazhi, Thrissur Cotton Mills at Nattika, Rajgopal Textiles at Athani, Sitaram Spinning and Weaving Mills, Thrissur and Vanaja Textiles at Kurichikkara. The mills namely the Cochin Hosieries Kuriachira, Thrissur and the Kunnath Textiles, Thrissur are engaged in the manufacture of hosiery products. The Madura Coasts at Koratty produce cotton sewing threads. The thread produced here is sold throughout the country. Sitaram Spinning and Weaving Mills the earliest textile mill in the district (1909) caught fire in December 1959. The mill started functioning later.

Tile Industry: The tile industry is the most important industry in the district employing the largest number of labourers. From a humble beginning early in this century, the industry had grown considerably in recent years. At present there are 160 tile factories in the district. Suitable clay required for the manufacture of tiles and bricks is found in Ollur, Pudukkad, Kaluvannor and Amballur which are the main centres of this industry.

Timber Industry: The timber industry of the district is of considerable importance. It had its beginnings in the first decade of this century when the first saw mill in the State was erected at Thrissur (1905) to convert teak and superior hard wood logs into slabs and other sizes.

Most Chalakudy, which are the most important timber marts in the district. In Chalakudy, Ollur and Thrissur, there are many saw mills with up-to-date plant and machinery.

Soap Manufacture: Soap manufacture is one of the flourishing industries of the district. It is mainly located in Irinjalakuda and Thrissur town.

Canning Industry: This is an industry that has recently sprung up and has immense prospects for development. The first unit of the Canning industries (Cochin) was started in Thrissur in 1947. The other two units are Darlco Cannings and Kayee Plantation Cannings, both situated at Thrissur r. Pineapple slices, Pineapple juice, tit bits, jams, squashes, syrups, jellies and marmalades are some of the products of these units. A canning industrial unit is being established at Nadathara by the Thrissur Fruits and Vegetables Marketing Society and it is going on very successfully.

Diary Unit: There is a diary unit at Ramavarmapuram in the public sector.

Chemicals: There are five units engaged in the manufacturers of chemicals Pharmaceutical products like elixirs, syrups, vitamin tablets, transfusion bottles, etc. In addition, some of the units manufacture commercial products like ink, paints, and varnish.

Oil Mills: Oil Mills are found in all parts of the district. Coconut oil is the most important product of these Mills. For a long time, the extraction of oil from copra was a cottage industry. Oil is also extracted from lemon-grass, gingelly castor-seed, groundnut etc.

Printing: The printing industry is fairly well developed in the district. Modern methods and techniques in printing are available in the district.

Match Industry: Soft wood required in the manufacture of match sticks is obtained from the local forests. Veneers and splints are also made in the match factories Cottage Industry: Handloom weaving is a premier cottage industry of the district.

It was practiced mainly by hereditary weaving communities like the Challias, Chettiars, Mudalis and Mudaliars. Poomangalam and Aripalam in Mukundapuram taluk and Kuttamippli and Thiruvilwamal in Talappilli Taluk are well known weaving centres of the district.

Coir Industry: Coir manufacturing is one of the important cottage industries of the district. The kind of yarn produced in the district is known are Chittattukara, Kottapuram and Kodungallor. Superior varieties of the quality of yarn known as 'Parur Special' are also produced in these areas. Another variety of yarn manufactured in the district is the rope yarn and the main centres of production are Kandassankadavu and Manalur.

Curing of Arecanuts: Arecanuts have to be cured for the market. Arecanut preparation is a seasonal industry of some importance. In the taluks of Talappilli, Thrissur and Mukundapuram hundreds of men and women are engaged in this occupation form September to January.

Cashew Industry: Thrissur district was the largest producer of cashew nuts next only to Kollam district in the State.

Grass mat and basket manufacture: The industry is an ancient one and products of this industry are widely used in Kerala. Very beautiful mats, either plain of with excellent designs are made.

Beedi Making Units: There are three Beedi making units at Ancheri, Chavakkad and Vadanappilly which are run by primary Beedi Co-operative societies under central Kairali Beedi co-operatives of which the headquarters at Shornoor.

Leather Industry: The manufacture of chappals, shoes suit cases and hand bags out of leather is an important industry of the district. Tanned leather is mostly procured from outside the district. Work in leather is the hereditary occupation of the "Chakkilins" or "Tholkollans", who are scattered in all parts of the district.

Thrissur is one of the most important centres of production of leather articles. The Foot Wear Service Centre at West Fort under the Ministry of Industries, Govt. of India imparts training in Shoemaking.

Engineering Workshops and Foundaries: Small smiths mending agricultural implements are found in rural areas. Repair shops have sprung up in towns. There are some umbrella manufacturing factories also in the district.

INDUSTRIAL ARTS (HANDICRAFTS)

Bell-metal Industry: Thrissur district is the largest producer of bell-metal articles in the State. The industry is monopolised by two castes - *Moosaries* and *Kammalas*. The main centres of production are Kadavallur, Kunnamkulam, Thrissur and Irinjalakuda. "Deepastambhams" and a few other articles are highly appreciated and there is a great demand for them in North India. A Bell Metal Workers Cooperative Society is successfully conducting large scale production of Bell Metal articles at Nadavaramba

Polishing of Imitation Stones: Thrissur, Ollur and Pudukkad are the chief centres of the industry. After being polished and processed the stones are exported to foreign countries. Now the imitation diamond manufacturing workers have been brought under a Central Cooperative Society called Diamond India, Thrissur Wood Carving: Wood carving is an important handicraft of the district. Almost the entire carving is done by carpenters hailing from Viswakarma community.

The wood carvers of Cherpu seven miles from Thrissur, are well known. The figure of elephants made in this place have a wide reputation. Carvings out of buffalo horn are also made here.

The carving of Kathakali dance dolls is also a special feature of this district. With the increasing appreciation of the art of this district. With the increasing appreciation of the art of Kathakali, the demand for these carvings has also been increasing. Another important handi-craft is "Alavattam" (peak-cock feather fan) made at Kanimangalam in Thrissur.

PUBLIC HEALTH SERVICE

Two of the Ashta Vaidyas in the field of Ayurveda viz; Kuttenchery Manu Moose and Thalikkattu Moose belong to Thrissur district. The western system of medicine was introduced here in the early part of the 19th century. There are 122 Allopathy hospitals and 14 Ayurveda hospitals and a homeopathy hospital in the district. Ayurvedic and Homoeo systems of treatments are very popular in the district.

Nature cure methods attract a large section of the people to Thrissur district. Prakruthi Chikkilsa Sahakaraana Sanitorium has been established to propagate Nature Cure among public. An Auyrveda Regional Research Institute under Government of India is functioning at Cheruthuruthy.

LOCATION & PHYSIOGRAPHY

The Kerala State lies along the south-west corner of Peninsular India, between 80 18' and 120 48' N latitude and 740 52' and 770 22'E longitude. The boundaries of the State are the Lakshadweep sea in the west, Tamilnadu in the south and east and Karnataka in the north. The State has an area of 38,863 km2, which is about 1.18 percent of the total area of the country and is administratively divided into 14 districts. Due to the long tract of Western Ghats along the eastern side and Arabian sea along the western side, the physiography of the State is highly diversified.

The State has a complex topography with mountains, valleys, ridges and scarps. The altitude varies from sea level to 2695 m above msl. Based on the altitude, the land is divided into high ranges (above 750 m asl; highlands (between 75-750 m asl); midland (between 7.5-75 m asl) and lowlands (below 7.5 m asl).

The highlands with an average height of 900 m have several peaks over 1,800 m and constitute about 43 per cent of the land area followed by midland (42 percent); high ranges (15 per cent) and lowland (10 per cent). A narrow strip of land bordering the sea constitutes the low land area of the State and this region holds the back waters and estuaries. Mangroves and coastal vegetation are confined to this region. Parallel to the coastal strip, there is wider more or less undulating midland zone.

Most of the human activities and agricultural settlements are concentrated in this region. The natural vegetation is rather scanty and occurring as small refugees. These two regions constitute the major human habitats in the State. Wider eastern highland region constitutes the important region with regard to the Biodiversity.

This region is highly undulating and has a complex geography compared to the other zones. These mountain ridges are continuous from north to south except the 30 km wide gap in the Palakkad district. These mountain chains influence the climate of the State to a greater extent.

PHYTOGEOGRAPHY

The Malabar coast, well known in the history of plant resource studies in Asia is remarkable for the luxuriant growth of tropical forests. Hooker (1907) classified the botanical regions of the erstwhile British India in to 9 regions based on species distribution.

The Malabar region is one among them consisting of Western Ghats and the west coasts. Clarke (1898) proposed 11 phytogeographical provinces for the British India. He also recognised the Western Ghats as a separate province 'Malabarica'.

Prain (1903) classified the phytogeograpic region based on moisture regimes into 6 regions. According to him 'India Aquosa' comprises the tropical rainforests along the Western Ghats. Chatterjee (1940) based on endemism among Dicotyledonous plants of the India-Burma region, recognised 10 botanical regions. He treated Malabar as a botanical region with a high percentage of endemism. All the above classifications recognised the Malabar region of the Western Ghats as a distinct phytogeographic region.

Because of the past cretaceous shield of Gondwana Kingdom, the floristic elements in the Peninsular India, especially the Western Ghats flora, shows differential similarity in the distribution of families and genera between sister geographical regions such as Malaysian Islands, Madagascar, Australia, S. America and Africa.

Among these countries Malaysian region is more similar to the peninsular Indian Flora. The polyphyletic origin of the taxa within this continents is widely accepted. Also the flora and fauna of this region are influenced by Pleistocene glaciations.

Homology in the evolution and geographical similarity also influence the development of the flora. In a more restricted sense the flora shows resemblance with Sri Lankan, north-eastern and lower Himalayas. The distributional records of several taxa strengthen this belief. Many taxa which were considered as endemic to Sri Lanka, north eastern India and Himalaya are reported from the Southern Western Ghats, especially from Kerala part. Western Ghats and Sri Lanka are more similar than other parts of the country. As a whole, the distribution of many groups of plants is restricted to this smaller bio-geographic zone. The peculiar endemic flora of this region represents the remnants of older flora and generally termed as paleo-endemics. About 30 percent of the Flora of Kerala are Peninsular Indian endemics. Because of the relict charactors, they are very sensitive to changes in the environment. Anthropogenic interference makes the situation more severe.

4

Topography of Thrissur Forests

Kerala is a state located on the South West edge of India. It is a narrow strip of land and lies between the Western Ghats mountain range on the Eastern side and the Arabian Sea on the Western.

Area	:	38,863 Sq.Km
Districts	:	14
Capital	:	Thiruvananthapuram (Trivandrum)
Language	:	Malayalam, Hindi, English
Time	:	Time GMT +5.30 hrs
Municipalities	:	60 (including townships)
Rivers	:	44
Longest River	:	Periyar (244 Km)
Highest Mountain	:	Anamudi (2695 Metres)
Climate	:	February to May (Max. 33°C Min. 24°C) Monsoon - June to September (Max. 28°C Min. 22°C) Winter - October to September (Max. 32°C Min. 22°C) September to May is the best time for visit
C M.L.A's	:	141
M.P's	:	(Lok Sabha) - 20
M.P's	:	(Rajya Sabha) - 9

The Kerala state is lined by coconut trees and beautiful beaches. The leaves of the coconut tree provide shade to almost the entire State from the tropical sun. Tourists can have a wonderful time riding on small boats through the beautiful backwater lagoons or canwatch elephants roaming in the wildlife sanctuaries.

Kerala has a rich tradition. History says that the Phoenicians, Romans, Arabs, and the Chinese have sailed to Kerala to trade for spices, wood and ivory. It was Vasco Da Gama who discovered India by landing to Kerala. The influence of Chinese articles such as Chinese fish in Kerala has started since their arrival to Kerala. The history of Kerala is a combination of old tradition with new values.

TOPOGRAPHY AND CLIMATE

Area and population: It is bounded on the north by Palakkad district, on the east by Palakkad district and Coimbatore district of Tamil Nadu, on the south by Ernakulam and Idukki districts, and on the west by the Arabian Sea. The area of the district is 3032 sq. km., while the population is 2,975,440 according to 2001 census.

Natural Divisions: Descending from the heights of the Western Ghats in the east, the land slopes towards the west forming three distinct natural divisions - the highlands, the plains and the sea board .

River System: The Periyar, the Chalakudy, the Karuvannur, and the Ponnani (Bharatha Puzha) are the main river systems in the district. They take their origin from the mountains on the east, and flow westward and discharge into the Arabian Sea. There are a number of tributaries also joining these main rivers. Climate: The district has a tropical humid climate with an oppressive hot season and plentiful and seasonal rainfall. The hot season from March to May is followed by the South West Monsoon season from June to September. The period from December to February is the North East Monsoon season, although the rain stop by the end of December and the rest of the period is generally dry.

PHYSIOGRAPHY

Around 48% of the total land is occupied by the Western Ghats which is also called as the Sahyadri. The highland rise up to an average height of 900m. There are numerous peaks in this Ghats which are over a height of 1800m. The total land covered is estimated to be around 18650 sq.km.

This area is ideal for cultivating crops such as tea, coffee, rubber and spices. This area is popularly called as the Cardamom Hills. It is called so because these regions yield cardamom in large quantities. The highest peak in the Western Ghats is in Munnar. The peak is called as Anaimudi with a height of 2694 meters. Many rivers originate from the Western Ghats.

The Midlands are made up of hills and valleys and they lie between the mountains and the lands. These midlands cover around 40% of the land and are estimated to be an area of 16200 sq.km. It is in this area that cultivation is mainly done. Various varieties of crops are grown here.

The remaining land which is about 4000 sq.km is covered by the lowlands which are also called as the Coastal Area. There are a large number of lagoons which are known by the names kayels, backwaters and coastal areas of the Arabian sea. The soil is very fertile and paddy is mostly grown in this area. Kerala is known as the land of coconuts and rice. In Kuttanad region the cultivation is done below the sea level. It is one among the few places in India where cultivation is done below sea level. The main means of transportation in these areas is through water.

RIVERS, LAKES AND BACKWATERS OF KERALA

Kerala has abundant water sources. Even though Kerala is small in size, there are 44 rivers, out of which 41 flow towards the west and the remaining 3 flow towards the east. Besides these main rivers there are also many tributaries and distributaries, numerous streams and brooks. All these water sources make Kerala a fertile land and is the main reason for its greenery. It also serves as inland waterways. Moreover there are countless lakes and backwaters which give extra beauty to the land. The largest lakeis the Vembanadu lake having an area of 260 sq.km and the largest fresh water lake is the Shastamkotta lake.

KERALA BACKWATERS

The Kerala backwaters are a chain of brackish lagoons and lakes lying parallel to the Arabian Sea coast (known as the Malabar Coast) of Kerala state in southern India. The network includes five large lakes linked by canals, both manmade and natural, fed by 38 rivers, and extending virtually half the length of Kerala state. The backwaters were formed by the action of waves and shore currents creating lowbarrier islands across the mouths of the many rivers flowing down from the Western Ghats range.

The Kerala Backwaters are a network of interconnected canals, rivers, lakes and inlets, a labyrinthine system formed by more than 900 km of waterways, and sometimes compared to the American Bayou. In the midst of this landscape there are a number of towns and cities, which serve as the starting and end points of backwater cruises. National Waterway 3 from Kollam to Kottapuram, covers a distance of 205 km and runs almost parallel to the coast line of southern Kerala facilitating both cargo movement and backwater tourism. The important rivers from north to south are; Valapattanam river (110 km.), Chaliar (69 km.), Kadalundipuzha (130 km.), Bharathapuzha (209 km.), Chalakudy river (130 km.), Periyar (244 km), Pamba (176 km), Achancoil (128 km.) and Kalladayar (121 km.). Other than these, there are 35 more small rivers and rivulets flowing down from the Ghats. Most of these rivers are navigable up to the midland region, in country crafts.

Ashtamudi Lake is the most visited of the lakes, covering an area of 200 km^2, and located in Kollam. The lake has a large network of canals that meander through the town. Ashtamudi is also India's most preserved lake

The backwaters have a unique ecosystem - freshwater from the rivers meets the seawater from the Arabian Sea. In certain areas, a barrage has been built near Neendakara Kollam, salt water from the sea is prevented from entering the deep inside, keeping the fresh water intact. Such fresh water is extensively used for irrigation purposes.

Many unique species of aquatic life including crabs, frogs and mudskippers, water birds such as terns, kingfishers, darters andcormorants, and animals such as otters and turtles live in and alongside the backwaters. Palm trees, pandanus shrubs, various leafy plants and bushes grow alongside the backwaters, providing a green hue to the surrounding landscape.

TOURISM

Kerala was placed among the '50 destinations of a lifetime' by National Geographic Traveler in a special collectors' issue released just before the turn of the millennium. House boat and backwater resort tourism in Kollam leads the Kerala Tourism to glory

House boat

The kettuvallams (Kerala houseboats) in the backwaters are one of the prominent tourist attractions in Kerala. More than 2000 kettuvallams ply the backwaters, 120 of them in Alappuzha. Kerala government has classified the tourist houseboats as Platinum, Gold and silver.

The kettuvallams were traditionally used as grain barges, to transport the rice harvested in the fertile fields alongside the backwaters. Thatched roof covers over wooden hulls, 100 feet (30 m) in length, provided protection from the elements. At some point in time the boats were used as living quarters by the royalty. Converted to accommodate tourists, the houseboats have become floating cottages having a sleeping area, with western-style toilets, a dining area and a sit out on the deck. Most tourists spend the night on a house boat. Food is cooked on board by the accompanying staff – mostly having a flavour of Kerala. The houseboats are of various patterns and can be hired as per the size of the family or visiting group. The living-dining room is usually open on at least three sides providing a grand view of the surroundings, including other boats, throughout the day when it is on the move. It is brought to a standstill at times of taking food and at night. After sunset, the boat crew provide burning coils to drive away mosquitoes. Ketuvallams are motorised but generally proceed at a slow speed for smooth travel. All ketuvallams have a generator and most bedrooms are air-conditioned. At times, as per demand of customers, electricity is switched off and lanterns are provided to create a rural setting

Now Ashtamudi Lake, Kollam has become the most visited place in Kerala While many ketuvalloms take tourists from a particular point and bring them back to around the same point next morning there are some specific cruises mostly in the Ashtamudi area, such as the one night cruise to Thotapally via Punnamada Lake two nights cruise to Alumkavadi, one night cruise from Alappuzha to Kidangara, and one night cruise to Mankotta. There are numerous such cruises.

Beypore, located 10 km south of Kozhikode at the mouth of the Chaliyar River, is a famous fishing harbour, port and boat building centre. Beypore has a 1,500 year-tradition of boatbuilding. The skill of the local shipwrights and boat builders are widely sought after. There is a houseboat-building yard at Alumkadavu, in Ashtamudi Kayal near Kollam.

Ferry services

Regular ferry services connect most locations on both banks of the backwaters. The Kerala State Water Transport Departmentoperates ferries for passengers as well as tourists. It is the cheapest mode of transport through the backwaters.

Resorts

Ashtamudi lake, which was a sleepy destination for years, has been transformed into a busy tourist destination with plush resorts around the Lake and the backwaters.

Impact on eco-system

The unregulated proliferation of motorised houseboats in the lakes and backwaters have raised concerns regarding the adverse impact of pollution from diesel engines and outboard motors on the fragile ecosystem.

ECONOMIC SIGNIFICANCE

Connected by artificial canals, the backwaters form an economical means of transport, and a large local trade is carried on by inland navigation. Fishing, along with fish curing is an important industry.

Kerala backwaters have been used for centuries by the local people for transportation, fishing and agriculture. It has supported the efforts of the local people to earn a livelihood. In more recent times, agricultural efforts have been strengthened with reclamation of some backwater lands for rice growing, particularly in the Kollam area. Boat making has been a traditional craft, so has been the coir industry.

Kollam region is crisscrossed with waterways that run alongside extensive paddy fields, as well as fields of cassava, banana and yam. The crops are grown on the low-lying ground and irrigated with fresh water from canal and waterways connected to Ashtamudi lake. The area is similar to the dikes of the Netherlands where land has been reclaimed from the sea and crops are grown.

ECOLOGICAL SIGNIFICANCE

Ashtamudi Wetland is included in the list of wetlands of international importance, as defined by the Ramsar Convention for the conservation and sustainable utilization of wetlands.

Boat races

Chundan vallams or snake boats are narrow boats over 100 feet (30 m) long, with a raised prow that stands 10 feet (3.0 m) above water and resembles the hood of a snake. Traditionally these were used by local rulers to transport soldiers during waterfront wars. In modern times, it has spawned a new sport

– the Vallam Kali (boat race). Each chundan vallam accommodates about a hundred muscular oarsmen. Boat races are occasions of great excitement and entertainment with thousands gathered on the banks to watch and cheer. Most of these races are held in the Kollam Region

The boat races starts with Champakulam Moolam Boat Race which is held on the Pamba River in the village Champakulam on Moolam day (according to the Malayalam Era M.E) of the Malayalam month Midhunam, the day of the installation of the deity at the Ambalappuzha Sree Krishna Temple. Very interesting stories lie behind the origin of Moolam Boat Race.

When Jawaharlal Nehru visited Kerala in 1952, four traditional chundan valloms went to receive him. A snake boat race was organised for him. He was so impressed that when he went back to Delhi, he sent back a gleaming silver trophy for a boat race. Even today, the 1.5 km Nehru Trophy Boat Race is the most prestigious. The Thazhathangadi boat race held every year on Meenachil river, at Thazhathangadi, Kottayam is one of the oldest and popular boat races in the state. Other renowned boat races are: Indira Gandhi Boat Race, Champakulam Moolam Boat Race, Aranmula Uthrattadi Vallamkali, Payippad Jalotsavam, kallada Boat Race andKumarakom Boat Race.

BACKWATER REGIONS

Kollam

Kollam (earlier known as Quilon) was one of the leading trade centres of the ancient world, eulogised by travellers such as Ibn Battutaand Marco Polo. It is also the starting point of the backwater waterways. The Ashtamudi Kayal, known as the gateway to the backwaters, covers about 30 per cent of Kollam. Sasthamcotta Kayal, the large fresh water lake is 28.5 km from Kollam city.

The 8 hours boat ride from Kollam to Alappuzha is the longest cruise in Kerala and is delightful ride with lotuses and water lilies all around. The historic Thangasseri Fort is near Kollam, which is situated 71 km north of Thiruvananathapuram.

Islands of Kollam

Islands are the eye-catching factors as well as the beauty of Lake Ashtamudi, Kollam. Most of these islands are potential tourism spots in the state. Even Indian Railways also planning to develop one of the islands in Kollam for a tourism project. There are big as well as small islands which are inhabited and uninhabited by human beings. There are more than 15 islands in Ashtamudi Lake. The important islands in Kollam are:

- Munroe Island
- Chavara Thekkumbhagom
- St. Sebastian Island

- San Thome Island
- Our Lady of Fatima Island
- Pezhumthuruth
- Kakkathuruth
- Pattamthuruth
- Paliyanthuruthu (*Palliyamthuruthu*)
- Neettum thuruth
- Puthenthuruth
- Poothuruth
- Pannaykkathuruth
- Veluthuruth
- Neeleswaram thuruth

Paravur

Paravur is one of the wonderful backwater destinations and a Municipal town situated in the Kollam district of Kerala state. The beautiful lakes and sea coast in Paravur attract visitors and foreigners. The main attraction is the interconnection of Paravur Kayal lake (part of a system of back waters) and the Arabian sea. There is a coastal road in between Paravur & Kollam. This road is starting in the western end of Paravur town, passing through the land between the banks of Arabian sea and Paravur Kayal(lake). The 14 km long journey through this particular road will be one of the amazing experience for the travelers. 3 Resorts are there in Paravur. Among them, one is situated in the middle of beautiful Paravur lake.

Estuaries of Paravur

Paravur Estuaries are the most scenic & eye-catching estuaries in India, lies near to the South-Western coast of Kollam. The place is world famous for its natural beauties, backwater locations, white-sand beaches and concentration of temples in each and every square kilometer. The peninsula of Paravur is one of the most visited one in Kollam district. Both north and south tips of Paravur town is having peninsula and estuary. Pozhikara is at north and Thekkumbhagam is at south of Paravur. One more estuary mouth in Pozhikara, which is very close to Pozhikara Devi Temple, which has breached in 2014 under the supervision of Water Resources Department (WRD), after a long gap of 14 years.

Munroe island

Munroethuruth or Munroe Island is a place surrounded by Kallada River, Ashtamudi Lake and Sasthamkotta Lake in Kollam district, Munroe Island is a cluster of eight tiny islands, Blessed with a number of criss-cross canals and zigzag water channels, this Island plays a host to many migratory birds from various countries around the world. You can watch birds such as Kingfisher, Woodpecker, Egret, Bee-eater, Crow pheasant, and Paddy Birds. There is yet another rare chance to see the traditional Indian spice plants such as Pepper, Nutmeg and Cloves. The first community tourism programme in the State will start functioning from the MunroeThuruthu islands. Coir making is a home industry to almost all the village living people. It is very interesting to watch the coir making by the village ladies with the help of weaving Wheels. They make the coir ropes by hand. In addition to this, on the way, you can see the process of extracting coconut oil from the "copra" [dried coconut]. Among the routine traditional engagements, duck, poultry farm and prawn breeding are common in all houses.

Kasargod

Kasargod in north Kerala is a backwater destination, known for rice cultivation, coir processing and lovely landscape, it has the sea to the west and the Western Ghats to the north and east. Cruise options are Chandragiri and Valiyaparamba near Kavvayi Backwater. Chandragiri is situated 4 km to the southeast of Kasargod town and takes tourists to the historic Chandragiri fort. Valiyaparamba is a scenic backwater stretch near Kasargod. Four rivers flow into the backwaters near Kasargod and there are many small islands along these backwater stretches, where birds can be seen.

Thiruvallam

Thiruvallam backwaters are just 6 km from Thiruvananthapuram, the state capital. Known for its canoe rides Thiruvallam is becoming increasingly popular with tourists. Two rivers, the Killi and the Karamana come together at Thiruvallam. Not far from Thiruvallam is the Veli Lagoon, where there are facilities for water sports, a waterfront park and a floating bridge. The Akkulam Boat club, which offers boating cruises on Akkulam Lake and a park for children, is also a popular tourist attraction near Thiruvallam.

Kozhikode

Kozhikode (also known as Calicut) has backwaters which are largely "unexplored" by tourist hordes. Elathur, the Canoly Canal and the Kallayi River are favourite haunts for boating and cruising. Korapuzha, the venue of the Korapuzha Jalotsavam is fast becoming a popular water sport destination.

Alappuzha

There are other backwaters in Alleppey also.

RIVERS AND LAKES IN KERALA

Kerala, the South Western state of Indian Peninsula can be rightly called the land of waterbodies with numerous majestic waterfalls and calm backwaters,

34 lakes with sparkling azure and emerald waters, and 49 rain-fed rivers with copious tributaries and distributaries flowing through the land.

Kerala is a natural paradise for nature admirers and there is no dearth of biodiversity in this state which boosts about not only its waterbodies but also other geographical features like hills, valleys and golden beaches.

Kerala is a natural paradise for nature admirers and there is no dearth of biodiversity in this state which boosts about not only its waterbodies but also other geographical features like hills, valleys and golden beaches.

Backwaters: The Backwaters consists of lakes and coves of oceans which straggle out in the land. The largest backwater is the Vembanad Lake which stretches out into the Arabian Sea at Cochin Port. The other important backwaters are Anjingo, Veli, Edava, Kadinakulam, Madayara, Kayamkulam, Paravoor, Kodungallur, Ashtamudi and Chetwa.

Lakes: The34 lakes in Kerala can be classified geographically into three categories. The first kind moves parallel to the riverbank and is edged by the sandbank. The second type has land at its frontage and the third type flows almost at right angles to the bank of the river. A large number of these lakes form superb destinations for backwater tours in Kerala. The still waters of the serene lakes offer a feeling of tranquility amidst nature and one seems to get lost in the world of his own thoughts. The most significant and famous lakes are -

Sasthamkotta Lake - Situated 19 km from the Kollam town, the Sasthamkotta Lake is one of its kind in the state of Kerala. It is a freshwater lake spread over a huge area of 375 hectares. The highlight of this lake is that it neither gets frozen during the winter season nor does it dries up in the summers. Throughout the year, the freshwater of this lake provides clean drinking water for around 10 million people. For tourists, it is not just the scenic beauty and tranquility of the lake that is attractive; rather there is something else as well. An ancient Sastha temple stands on its shore which is major draw for pilgrims to Kerala

Vembanad Lake - This freshwater lake is situated 12 km west of Kottayam and is internally connected with a number of rivers and canals. Tourists can have loads of excitement by cruising in the lake and enjoying the blissful beauty of the lake.

The Ashtamudi Lake - This is another significant lake in Kerala that exhibits an idyllic beauty that is very pleasing to our eyes. The Ashtamudhi Lake has eight channels and hence it has this name. The lake looks unique with the Cheena Vala, the famous Chinese fishing net in its background. You can head towards the Kollam backwaters by sailing dreamily in a boat as you will be enchanted by its spell-binding beauty.

Akkulam Lake - This Lake is situated at a little distance from Thiruvananthapuram. This lake is an ideal site for a private picnic and there is also a children's park adjacent to it. Tourists love lounging at the park and sitting lazily by the cool waters.

Pookot Lakes - This pristine lake, bounded by gigantic mountains is located at a distance of 3 km from south of Vythiri town.. Boating in the Pookot Lake is a delightful experience. Besides there is also a freshwater aquarium with a large number of multi-colored fish, a children's park and a shopping centre for handiworks and spices.

River: There are 49 rivers flowing in the state of Kerala, among them 46 flows to the west and the other 3 flow to the east. They evolve from Western Ghats and rush towards the west into the Arabian Sea. All these rivers are rain-fed, non-perrenial Rivers that are short and are transformed into streams during summer season. The most significant rivers from north to south are-

- Valapattanam River(110 kms.)
- Chaliar (69 kms.)
- Kadalundipuzha(130 kms.)
- Bharathapuzha (209 kms.)
- Chalakudy River (130 kms.)
- Periyar (244 kms)
- Pamba (176 kms)
- Achancoil(128 kms.)
- Kalladayar(121 kms.)

Besides these, there are more 35 small rivulets and rivers that flow down from the Western Ghats and most of these rivers are crossable up to the midland region and thus provide the means of economic transport for boats, ferries and etc.

The existence of these rivers in abundant numbers has made Kerala a resourceful state and Kerala has been able to harness a lot of Hydro- electric power through its waterbodies.

BACKWATER DESTINATIONS

Kerala waterways

The entire state of Kerala is blessed by nature with plenty of rivers, lakes, ponds, lagoons, and backwater stretches.The waterways of Kerala play a main role in the economy of the state,and they link remote villages and islands with the main land. The water ways provide a cheap and economical mode of transport. It is an incredibly different experience to cruise in the backwaters in country boats, absorbing the beauty of Kerala villages. Kerala is known for its panoramic backwater stretches, lush green paddy fields, highlands and beaches. The scenery flashes up vivid contrasts of the breath-taking green landscape and the deep blue warters. Boat rides, houseboat cruises and holiday packages are available for the tourists at nominal rates.The house boat cruises along the backwaters of Kerala have become a rage among Indians as well as international tourists. Kerala's palm-lined backwaters are actually inland lakes linked by a network of canals. With 41 West-flowing rivers, the backwaters stretch up to about 1,500 km. The backwater routes date back over the centuries and have been long used for all transition needs, in particular trade in coconut, rubber, rice and spices. Today, these waterways connect distant villages and islands to the mainland and are nerve centers of the coastal area of Kerala.

Backwater Tours/Cruises in Kerala

Backwater tour/cruise in Kerala is a great way of spending your holidays in the state.The splendid sunsets, the waxing moonlights, the shoals of ducks, the pulsating palms and the wonderful waves beckon the tourist to take a dream trip down the backwater destinations in kaleidoscopic Kerala.

The major backwater stretch is in Kottayam district, where a network of rivers and canals empty into the great expanse of water called Vembanad Lake . Located at Kumarakom 16 km from Kottayam town, Vembanad Lake , an enchanting picnic spot and a fast developing backwater tourism destination, provides boating, fishing and sight seeing opportunities that are truly memorable. Kerala backwater cruises complete the tour to southern Kerala. Kerala backwaters take you through the winding waterways of God's Own Country, unveiling the colourful culture and customs of remote Kerala. The coastline of Kerala is dotted with these delightful backwater destinations.

Backwater Destinations - Kasaragod District

The northernmost district of Kerala, Kasaragod is situated on the sea coast bordered by hilly Kodagu and Mangalore districts of Karnataka on the east and north.Fishing is a prime source of livelihood along with the coir and handloom industries. With its unique natural and cultural attractions , Kasaragod is known as the land of Gods forts, rivers, hills and beautiful beaches.The Bekal fort, which stands on a 35 acre headland that runs into the Arabian sea , is the largest and the best preserved fort in the State.

Chandragiri

Situated 4 kms southeast from Kasaragod town on the Chandragiri river,this place is known for its large 17th century fort built by Sivappa Naik of Bedanore. The Chandragiri fort, one among a chain of forts built by the same ruler, offers a breathtaking view of the river and the Arabian sea . It is a vantage point to watch the sunset.

Chandragiri Cruises

Boat trips to the nearby islands and palm groves are available.Boarding point: Chandragiri bridge.

Valiyaparamba

Situated 30 km from Bekal fort,this is perhaps the most scenic backwater stretch in Kerala .It is fed by four rivers and dotted with numerous little islands.

Valiyaparamba Cruises

Valiyaparamba is fast turning into a much favoured backwater resort that offers enchanting boat cruises.

Backwater Destinations - Kozhikode District

Once the capital of the Zamorins, a prominent trade and commerce centre and the land of the Malabar Mahotsav, Kozhikode was the most important region of Malabar in the days gone by.

Today, ancient monuments,the lush green countryside, serene beaches, historic sites, wildlife sanctuaries, rivers, hills, a unique culture and a warm, friendly ambience make this district a much sought after tourist destination.

Unexplored and unspoiled, the backwaters of Kozhikode hold great promises of enchanting holiday options. Elathur, the Canoli canal and Kallai river are favourite haunts for boating. Kadalundi with its beautiful bird sanctuary is a charming site.

Korapuzha, the venue of Korapuzha Jalotsavam is fast becoming a popular water sport destination.

Backwater Destinations - Ernakulam District

More popular as the Queen of the Arabian Sea , Kochi is a cluster of islands on the vast expanse of the Vembanad Lake.Some of these picturesque islands are Bolgatty,Vypeen , Gundu and Vallarpadam.

The lake opens out into the Arabian Sea here to form one of the finest natural harbours in the world. It is this natural advantage that has made Kochi, a fascinating blend of the cultures and influences of explorers and traders who visited this wealthy land.

The Arabs, Chinese, Portugese, Dutch and the British have all left their mark here. The Jewish synagogue , the Dutch palace,the Chinese fishing net and other remnants of European and Asian architecture merge smoothly into the traditional fabric of this seaport city.

Backwater cruises conducted by KTDC

1. Boat cruise: (Daily two trips: 0900 - 1230 & 1400 - 1730hrs) from Kochi harbour to Willington island , Mattancherry palace, the Jewish synagogue , Fort Kochi and Bolghatty island.
2. Sunset Cruise (Daily trip: 1730 - 1900 hrs)
3. Three hour backwater village (daily) cruises which include visits to coir villages and coconut plantations.

Backwater Destinations - Kottayam District

Bordered by the lofty Western Ghats on the east, the Vembanad lake and the paddy fields of Kuttanad on the west, Kottayam is a land of unique characterestics.

Panaromic backwater stretches, lush paddy fields, highlands, extensive rubber plantations and the total literacy level achieved by the people have given this district the title: the land of letters, latex and lakes.

Kumarakom

A voyage towards the north of Alappuzha, about 10 Kms from Kottayam, takes you to the rich green, sleepy little village of Kumarakom on the Vembanad lake. An enchanting picnic spot and a fast developing backwater tourism destination, Kumarakom provides boating , fishing and sight-seeing experiences that are truly exhilarating. An exclusive attraction of this much sought after backwater resort is the Kumarakom bird sanctuary. The Kumarakom bird sanctuary is an orthithologist's paradise and a favourite haunt of migratory birds like the Siberian stork, egret, darter, heron and teal. A cruise along the Vembanad lake is the best way to view the bird life.

Backwater Destinations - Alappuzha District

With the Arabian sea on the west and a vast network of lakes, lagoons and waterways crisscrossing it, Alappuzha is a district of immense natural beauty. Referred to as the Venice of the east by travellers from across the world, this backwater town is also home to diverse animal and bird life.

By virtue of its proximity to the sea , the town has always enjoyed a unique place in the maritime history of Kerala. Today, Alappuzha has grown in importance as a backwater tourist centre, attracting several thousands of foreign tourists each year. Alappuzha is also famous for its boat races, houseboat holidays, beaches, marine products and coir industry. A singular characteristic of this land is the region called Kuttanad.

Kuttanad

Kuttanad, called the rice bowl of Kerala because of her wealth of paddy crops, is at the heart of the backwaters.The scenic countryside of Kuttanad with its shimmering waterways also has a rich crop of bananas, yams and casava which accompany the rice bowl as "side dishes". This is one of the places in the world where farming is done below sea level. Inland waterways which flow above the land level is an amazing feature of this unique place.

Pathiramanal

According to mythology, a young brahmin dived into Vembanad lake to perform his evening ablutions and water is said to have made way for land to rise from below, thus creating the enchanting island of Pathiramanal (sand of midnight). This little island on the backwaters, situated 14 km from Alappuzha, is a favourite haunt of hundreds of rare migratory birds from different parts of the world. The island lies between Thaneermukkom and Kumarakom and is accessible by boat.

Backwater Destinations - Kollam or Quilon District

Located 71 Kms to the north of Thiruvananthapuram, Kollam is another coastal district of Kerala.The district which is the centre of the country's cashew trading and processing industry also has some interesting historic remnants and a number of temples built in traditional ornate architechural style.

One of the oldest ports on the Malabar coast , Kollam was once the port of international spice trade . 30% of this historic town is covered by the renowned Ashtamudi lake, making Kollam the gateway to the magnificient backwaters of Kerala . The eight hour trip between Kollam and Alappuzha is the longest backwater cruise in Kerala

Alumkadavu

Alumkadavu, situated 23 kms from Kollam town,is halfway on the route to Alappuzha from Kollam. The quiet little backwater village in Karunagapally town is famous for its boat building yard. It is here that the gigantic kettuvallam(traditional cargo boats of rural Kerala) were built. Today the kettuvallams, which were long replaced by mordern means of transport, have been converted into full fledged houseboats.

Backwater Destinations - Thiruvanathapuram District

The southernmost district of the state, Thiruvanathapuram is bounded by the wooded highlands of the Western Ghats on the east and the north east , and the Arabian sea on the west.A long shoreline with secluded internationally renowned beaches, historic monuments, backwater stretches and a rich cultural heritage provide this district some of the most enchanting picnic spots.

FOREST

The total forest area in Kerala is estimated to be 11,125.59 sq.kms. There are also 12 Wild life Sanctuaries and 2 National Parks in these. The area under forest covers a total of 28.88 % of Kerala. The forest area is mainly in the Western Ghats. The forests in Kerala can be categorised into 5 main groups. They are Tropical Wet Evergreen Forests, Tropical Moist Deciduous Forests, Tropical Dry Deciduous Forests, Mountain Sub Tropical and Plantations.

FLORA AND FAUNA

The Western Ghats is one of the World's 18 hot spots of bio-diversity. It is rich in wide variety of rare flora and fauna. The rivers, mountain rain forests, tropical for ests, grasslands etc makes it suitable for a variety of plants and animals to inhabit these areas. The animal life includes elephant, tiger, sambar deers, cobras, hornbills and many others. A total of 25 % of nations 15000 plant species are found in Kerala.

The mountain ranges with thick evergreen forests afford ideal abode for various animals and game including diverse birds while the middle country with hills and low plateau, mostly cleared for cultivation and human habitation, still affords shelter and food for many of the smaller mammals, birds and reptiles and also many lower animals of diverse groups. The lowlands of the extreme west, bordering the coastline are dotted with backwaters and estuaries of rivers, all connected by an interesting system of canals forming a continuous waterway. Its waters abound in fish and afford feeding ground for many water birds, local and migrant, while the plains have rich fauna representing all groups. Among the mammals the Primates are represented by the langurs and monkeys. Coconut palm and paddy are mainly cultivated in the lowlands.

CLIMATE

Kerala lies near to the equator, but due to the presence of sea and the Western Ghats, Kerala enjoys an equable climate that changes little from season to season. The temperature usually varies from 27° to 32° C. The two main rainy seasons in Kerala are the Northeast Monsoon and the Southwest Monsoon. The annual average rainfall received in Kerala is 118 inches.

Extend

Kerala constitute just 1.03 % of India and has an area of 38,863 sq.km. The coastline extends to about 580 km. The width of this state varies from west to east where the maximum is 120 km and minimum is 30 km.

Neighbours

As Kerala is a coastal land it is bordered three sides by land and one side by the Arabian sea at the west. The neighbouring states of Kerala are Karnataka in the north and Tamil Nadu sharing the border with the rest of Kerala. So the whole of the western and southern frontiers of Kerala is bordered by Tamil Nadu.

Mineral Resources

The coastal regions of Kerala are rich in variety of minerals. The mineral ores present in Kerala are silica, quartz, bauxite and sillimanite. Moreover, the best and finest type of China Clay is available in this state.

Natural Resources

The greatest resource of Kerala is her natural beauty with her blue mist - capped mountains, undulating hills and valleys of a thousand shades of green, the blue spread of an endless stretch of lagoons and long seacoast blessed by world famous beaches coupled with an amiable people of rich culture, and art forms which all together earned for her the sobriquet God's own country making her one of the biggest tourist attractions of the world.

Kerala's forests are rich with the pristine silent valley and Agasthyavanam biospheres home to a plethora of rare species of plants and animals making them one of the seven biological hot spots of the world. Kerala ranks 14th among all the States/Union Territories in respect of the geographical area under forest cover which as per 2003 assessment of FSI was 15577 sq.km (dense forest 334 sq.km moderate dense forest 9294 sq.km. and open forest 5949 sq.km.) Forest plantation constitutes major source of raw materials to the forest based industries. Teak is the major species planted in 75767 ha. followed by eucalyptus 14274 ha. softwood species 28832 ha. bamboo and reeds 5912 ha. Plantation crops 1712 ha.

Kerala has 41 west flowing rivers and three east-flowing rivers namely Pambar, Bhavani and Kabani. The biggest river system among the west flowing rivers is Bharatapuzha with most of its water impounded in the Parambikulam-Aliyar dam providing irrigation to Tamilnadu and its tributary Siruvani providing drinking water to Coimbatore. The major portion of water of the second biggest river system Periyar is impounded in the Mullaperiyar dam providing water for irrigation in the neighbouring districts of Tamil Nadu. Other important river systems are the Chaliyar, Pamba etc. Kerala also has constructed a number of irrigation dams like Malampuzha for irrigation purposes. There are 24 hydel projects owned by Kerala State Electricity Board with an installed capacity of 1883.60 M.W. Kerala has tremendous wind power with only a fraction of it tapped by the 2.25 MW wind farm at Kanjikode in Palakkad district and a 40 MW windfarm at Ramakkalmedu in Idukki district.

The water resources in the State are not abundant but over exploited. The drinking water projects of five corporations and most of the 53 municipalities are based on the water yielded by the west flowing rivers. Although Kerala gets a normal rain of 2360.6mm per year the topography of Kerala with undulating hills and valleys and deep rivers except the narrow strip of the coastal region allows poor retention of water. So rainwater retention structures and watershed management schemes are needed. In Kerala ground water occurs under phreatic, semi-confined and confined conditions. Ground water is largely concentrated in the sedimentary aquifers of the coastal regions. It is mainly tapped for drinking and to a lesser degree for irrigation. The ground water potential of Kerala is very low compared to that of other States. The replenishable ground water resource is only 6841 million cubic meters.

The State has all the requisite natural endowments for building a strong fisheries economy with 590 km. Coastal belt and inland water spread of over four lakh hectares. The exclusive economic zone lying adjacent to the coast is

36,000 sq.km. almost equivalent to the land area of the State. The State has a fresh water area of 158358 ha. consisting of reservoirs 42890 ha. private ponds 21986 ha. irrigation tanks 1847 ha. and the area of rivers 85000 ha. The brackish water resource consists of 65213 ha. of estuaries and backwaters and 12873 ha. of prawn filtration fields. The polders of Kuttanad have a water-spread area of 35,000 ha. Another 17000 ha. area of Kole lands of Thrissur is also ideal for aquaculture development activities.

Kerala is endowed with a number of deposits such as heavy mineral sand containing mainly thorium, titanium, lithium, zircon etc, china clay, iron ore, graphite and bauxite, silica sand, lignite, lime-shell, granite etc. However mineral sand and china clay contribute more than 90 percent of the total value of mineral production in the State.

WATER

From the point of view of water resources Kerala is having both abundance and scarecity. The average annual rainfall of the state is 3000mm, the bulk of which (70%) is received during the South-West monsoon which sets in by June and extends upto September. The state also get rains from the North-East monsoon during October to December.However the spatial and temporal distribution pattern is mainly responsible for the frequent floods and droughts in Kerala. The average annual rainfall in the lowland of Kerala ranges from 900mm in the south to 3500mm in the north. In the midland, annual rainfall ranges from 1400mm in the south to about 6000mm in the north. In the highland, annual rainfall varies from 2500mm in the south to about 6000mm in the north. Kerala has got 41 west-flowing and 3 east-flowing were originating from the Western Ghats. The total annual yield of all these rivers together is 78.041 Million Cubic Meters (MCM) of which 70,323 MCM is in Kerala. The peculiarity of the rivers flowing across Kerala is short length of the river and the elevational difference between the high and the low land leading to quick flow of water collected from the river basin and quickly discharged into the Lakshsdweep sea, the state has not been able to utilise the river water sources to a major extent. The major portion of the runoff through the rivers takes place during the monsoon seasons. 67.29% of the surface water area of 3.61 lakh hectares is constituted by brackish water lakes, backwaters and estuaries.

On a rough estimate, the source wise dependence by rural households for domestic water supply dependent on traditional ground water systems is 80%, 10-15% use piped water supply systems, and 5% use traditional-surface and other systems.

Table. Fresh water availability in Kerala

Per capita water availability, litres/day			
Year	**Rain**	**Surface water**	**Ground water**
2001	9450	1022	590

Surface Water Resources

Kerala is rich with 44 rivers which together yield 70300Mm^{3}of water annually. However the total utilizable yield is estimated to be 42000Mm^3, only 60% of the annual yield. Kerala possess only four medium rivers and 40 minor rivers.

In the all India perspective the rivers of kerala are not so significant than even the largest of them cannot find a place among the major Indian rivers. With respect to the national norm Kerala does not have a single major river and has only four medium rivers. The combined discharge of these four rivers is less than half of that of river Krishna. The remaining fourty rivers are only minor ones, the combined discharge of all of them together is only about one-third of that of Godavari. western ghats from where the river originate is devoid of snow and therefore these river systems do not have the benefit of water supplied during the summer seasons as in the north Indian rivers.

Ground Water Resource of Kerala

Groundwater has been the mainstay for meeting the domestic needs of more than 80% of rural and 50% of urban population besides, fulfilling the irrigation needs of around 50% of irrigated agriculture. The ease and simplicity of its extraction has played an important role un its development. Recent the problems of decline in water table, contamination of groundwater, seawater intrusion etc. are being reported at many places.

Ground water fulfils the irrigation needs of around 50 percent of irrigated area. The total Annual ground water availability in Kerala State has been computed as 6.620 Billion Cubic Meter (BCM) and the net ground water availability in the entire state is 6.029 BCM. The rainfall recharge accounts for about 82 percent of the annual recharge. The annual ground water draft for all uses in the state is 2.809BCM. The net Ground water availability for future irrigation development in the state as in 2009 is of the order of 3.021 BCM. The overall stage of development of the State is 47 percent. The district wise stage of development is maximum in Kasaragod (71 percent) and minimum in Wayanad district (17 percent).

The net annual ground water availability for the state of Kerala during 2009 has reduced to 3.22 per cent compared with the data during 2004. The annual ground water draft for all uses has reduced by 3.80 per cent during the period. The net ground water availability for future irrigation development in the state as a whole shows a decline of 1.74 per cent in 2009 compared to 2004.

Hydrogeology

Kerala State is a narrow stretch of land covering 38863 sq.km area bordering theLakshadweep Sea on the western side and Tamil Nadu Karnataka Station the eastern side. The length of the State from north to south is 560km and the average width is 70km with a maximum of 125km. it lies between North latitudes 08 0 18' and 12 0 48' and east longitudes 74 0 52' and 77 0 22'.

Occurrence of Groundwater

Ground water occurs under phreatic, semi-confined and confined conditions in the above formations. The weathers crystallines, laterites and the alluvial formations from the major phreatic aquifers, whereas the deep fractures in the crystallines and the granular zones in the Tertiary sedimentary formations form the semi-confined and confined aquifers.Along the hill ranges, the crystalline rocks are covered by thin weathered zone. Thick zones of weathered crystallines are seen along midland region. The depth to water level in the weathered crystallines in the midland area ranges from 3 to 16mbgl. The midland area sustains medium capacity dug wells for irrigation. Mostly dug wells that can cater to domestic needs are feasible along topographic lows. Bore wells tapping deeper fractured aquifer are feasible along potential fractures in the midland and hill ranges. Potential fractures are seen down to 240m and the most productive zone is between 60 and 175m and the discharge of bore wells range between 36,000 and 1,25,000 lph.

Of the four Tertiary beds, the two beds viz. the Vaikom and Warkali beds are potential aquifers. The Alleppey beds at the bottom contains bhrackish water as inferred from electrical logs, whereas, the Quilon beds are poor aquifers. The Vaikom aquifer is seen all along the coast between Quilon and Ponnani and the piezometric surface ranges from 1 to 18 m above msl. The aquifer is extensively developed between Quilon and Kayamkulam. The aquifer contains fresh water south of Karuvatta in Alleppey district and also north of chellanum in Ernakumal district. The Warkalai aquifer is seen south of Cochin . The piezometric head in the aquifer varies from 2.6m above msl to 10m above msl. The aquifer is largely developed in and around Alleppey and in Kuttanad area.

Laterites are the most widely distributed lithological unit in the State and the thickness of this formation varies from a few meters to about 30m. The depth to water level in the formation ranges from less than a meter to 25 mbgl. Laterite forms potential aquifers along valleys and can sustain medium duty irrigation wells with the yields in the range of 0.5-6m^3 per day.

The alluvium forms potential aquifer along the coastal plains and ground water occurs under phreatic and semi-confined conditions in this aquifer. The thickness of this formation varies from few meters to above 100m and the depth to water level ranges from less than a meter to 6m bgl. Filter point wells are feasible wherever the saturated thickness exceeds 5m.

Depth to Water Levels

The depth to water level was monitored from 866 NHS distributed throughout the State during the months of January, April, August, November, 2001. The water level measured during the month of April is considered to be pre-monsoon water level and the data of November months are considered to be post monsoon water level.

The depth to water level mostly depends upon the hydro geological conditions of the area as well as topography, rainfall pattern, etc. In coastal plains the depth to water level is generally restricted to 6 m bgl. In midland areas, where the undulating topography is seen, the depth to water level generally varies from near ground level to 15m bgl. The variation is mostly due to topographical variations, thickness of lateritic overburden etc. In select areas where laterites are underlain by sedimentary aquifers of Tertiary age, the water level goes very deep, even to the extent of 55.0m.bgl.

Ground Water Potential of Kerala

The ground water potential of Kerala is very low as compared to that of many other states in the country. The estimated ground water balance is 5590Mm^3. Dug wells are the major ground water extraction structure in Kerala. The dug wells have a maximum depth of about 10 to 15 meters and have a diameter of about 1 to 2 meters in coastal region and 2 to 6 meters in the midland and high land. The open well density in Kerala is perhaps the highest in the country – 200 wells per sq.km in the coastal region, 150 wells per sq.km in the midland and 70 wells per sq.km in the high land. The ground water withdrawal is estimated as 980Mm^3 and the State Ground Water Department calculate the effective recharge as 8134 sq Mm^3.The ground water level receding drastically during the summer months and drying up of wells are common features of the ground water levels in many parts of Kerala.

The ground water replenishment and hence the levels depends also on the geo-morphological, physical and chemical properties of the soil in general, The depth of water level in Kerala state varies from few cm bgl to 56 M bgl and most of the area fall under 0-20 M bgl. The depth of the water level in the weathered crystalline of midland areas in Kerala varies from 3- 16 M bgl. The midland area sustains medium capacity dugwells. Borewells tapping deeper fractured aquifer are feasible along potential features in the midland and hill ranges. Potential fractures are seen down to 240 M and the most productive zone is between 60 M and 175 M.

The discharge of borewells range between 3,600 Iph and 1,25,000 Iph. In laterites, which is the most widely distributed lithological area in the state having a thickness from a 3 M to 30 M, the depth of water level ranges from less than a meter to 25 M.bgl. Lateries from potential aquifer along valleys and can sustain wells with yields in the range of 0.5 M^3 to 6 M^3 per day. Along the coastal plains the ground water occurs at depth ranging from less than a meter to 6 M.bgl. Filter point wells are feasible wherever the saturated availability indicate that ground water depths are farthest for laterite regions and shallowest for coastal alluvium during all times of the year. The availability of the groundwater level between the post and ore monsoon levels varies widely. The water level fluctuations in the post monsoon and ore monsoon vary between coastal alluvium, river alluvium and valley hills.

Groundwater Management

The National Water Policy of the Government of India states that the non conventional method for utilization of water such as through artificial recharge to ground water and traditional water conservation practices like rainwater harvesting need to be practiced to increase the utilizable water resources. The rainwater harvesting can be effected by in-situ-Harvesting and artificial recharge to ground water is the process of diverting the surface water into suitable geological formation. The common structures are percolation tanks, khadins, check dam/Anicut, sub-surface dams and injection wells. The ground water storage is the best method for water harvesting as it not only involves filtration of surface but is also safe from evaporation losses, natural catastrophes etc.

Central Ground Water Board has implemented various artificial recharge schemes in Kerala like surface dykes, percolation tanks, and of top rainwater harvesting. Four sub-suface dams were constructed at Palghat district (Anaganadi, Bhabaji Nagar, Alanallur and Ottappalam), one at Ernakulam (Odakali), one at Kottayam (Neezhir) one at Quilon (Sandanadapuram) and two at Trivandrum district (Mampazhakara and Ayiolam). Central Ground Water Board has constructed two percolation tanks, one at Chirakulam of Kottayam district and another one at Kadapallam of Kasaragod district. Roof top rainwater harvesting schemes were implemented at two places viz. Ezhimala and Mayyilcolony of Kannur district. The artificial recharge structures have given satisfactory results and the groundwater condition in the area has improved considerably.

Rainwater harvesting is the viable solution in the monsoon rich state of Kerala. The common structures feasible for Kerala are sub-surface dykes, nala bunds, check dams. The traditional water conservation structures like natural ponds, reservoirs should be desilted and cleaned. Participatory watershed development programmes should be implemented in the State. Mass awareness programme on ground water conservation should be arranged at Panchayat level in all districts.

Other Resources

Apart from rivers and wells sources like tanks, ponds, springs and surangams are also use in Kerala for providing water for drinking as well as irrigation. It is estimated that Kerala has approximately 995tanks and ponds having more than 15000 Mm^3 summer storage. Natural springs occurring in the highland regions of Kerala state have the potential to be developed as good sources for drinking water supply and also for limited small scale irrigation, especially in remote and under developed areas. A total of 236 springs have been identified in the state. Kasaragode district in Northern Kerala has 510 special kind of water harvesting structure called Surangams which have >111pm discharge.

Table. Major Irrigation Projects and Irrigation Status of Kerala(2011)

No	Name of Project	District	Year of Start	Year of Completion	Expen-diture	Ayacut Net	in ha Gross
1	Neyyar	Thiruvananthap uram	1951	1976	461.00	15380	23470
2	Pampa	Pathanamthitta	1961	1994	5898.04	21135	48480
3	Periyar Valley	Ernakulam	1956	1994	8350.87	32800	78325
4	Chalakkudy	Thrissur	1949	1966	188.25	19696	27258
5	Vazhani	Thrissur	1951	1962	107.57	2113	4226
6	Cheerakuzhy	Thrissur	1957	1973	90.76	1619	1846
7	Malampuzha	Palakkad	1949	1966	580	21732	40208
8	Peechi	Thrissur	1947	1959	235	18623	23718
9	Mangalam	Palakkad	1953	1966	106	3639	6608
10	Wayalar	Palakkad	1956	1964	131.66	3844	6505
11	Gayathri	Palakkad	1956	1970	220	5466	10114
12	Pothundy	Palakkad	1958	1971	234.25	5466	10046
13	Chitturpuzha	Palakkad	1963	1994	2570.21	15700	29950
14	Kuttiady	Kozhikode	1962	1994	5072.69	14570	34710
15	Chimoni Mupli	Thrissur	1975	1996	5958	13000	26000

Ongoing Irrigation Projects						
				Estimate	Expected Net	Ayacut in ha. Gross
1.	Vamanapuram	Thiruvananthapuram	1981	3640	8057	16436
2.	Kallada	Kollam	1961	45780	61630	92800
3.	Thanneermukkam	Alappuzha	1975	1650	--	--
4.	Meenachi	Kottayam	1980	4956	9950	14510
5.	Moovattupuzha	Ernakulam	1974	8925	17737	34737
6.	Edamalayar	Ernakulam	1981	6940	14060	43190
7.	Kanjirapuzha	Palakkad	1961	7500	9713	21835
8.	Kuriyarkutty-	Palakkad	1978	6018	11736	23470
	Karappara	Palakkad	1975	5000	4347	8378
9.	Attappady valley	Palakkad	1998	--	1303	3997
10.	Thrithala (BCR)	Malappuram	1981	37800	73240	108035
11.	Chaliyar	Malappuram	1985	1765	3106	9659
12.	Chamravattom(BCR)	Wayand	1979	1798	2800	4800
13.	Banasurasagar	Wayanad	1975	4066	4650	9300
14.	Karapuzha	Kannur	1962	7736	11525	23050
15.	Pazhassi	Kasargod	1979	9885	13980	41760
16.	Kakkadavu					

The overall performance of the major and medium irrigation sector during the initial years was not encouraging. The cumulative area brought under irrigation through major and medium irrigation projects is 29346 hectares (gross). The details of the progress of implementation of ongoing projects as on March 2010 are given below.

Project-wise Details of Ongoing Projects

(` Lakh)

Sl. No.	Name of Project	Year of starting	Original estimate	Revised estimate	Year of revision	Cost escalation (%)	Expenditure upto March 2010	Target area to be irrigated (Ha)		Physical achievement as on 3/10 (Ha)	
								Net	Gross	Net	Gross
1	2	3	4	5	6	7	8	9	10	11	12
1	Muvattupuzha	1974	4808	79300	2008	1649	75644	19237	37737	15026	29452
2	Idamalayar	1981	904	60000	2008	6637	36842	14394	29036	-	-
3	Karapuzha	1979	760	49800	2008	6553	27513.65	5221	8721	-	-
4	Banasurasagar	1999	1137	12700	2008	1117	3501.28	2800	4740	-	-
5	Chamravattom	1983	7000	11417	2008	163	3965.48	3170	4344	-	-

Karapuzha Irrigation Project

Karapuzha Project is the first project for irrigation taken up in the Wayanad District of Kerala. The project is to construct an earthen dam with concrete spillway in Right bank at Vazhavatta across Karapuzha stream and the saddle dam at Pakkam, Cherupetta and Cheengeri to create a reservior of 76.50 Mm^3 storage capacities. The project was originally envisaged for Irrigation only and now it has turned to be a multipurpose project. The original estimate of the project was Rs.7.60 crores in 1979and the estimate as per the 2008 schedule of rates is Rs.498 crores. The cumulative expenditure up to March 2010 is Rs.275.14 crores. Major components under head works viz., (i) Earth dam and saddle dams and (ii) Spillway are already completed. Work of stilling Basin and Energy Dissipating Chamber connection structure to RBC from diversion chamber is in progress. Raising of roads completed to the extent of 90 per cent. As regards Left Bank canal, it is in the nearly completion stage (99%). Seventy five per cent of the work in respect of Padinjareveedu Branch canal of LBC is completed, investigation of Thondippally Branch and Kuttoor Branch of Left bank canal is completed for which land acquisition is in process.

Idamalayar Irrigation Project

The Idamalayar Irrigation Project is a diversion scheme intended to irrigate an extend of 14394 ha. of wet and dry lands and the Cultivable Command Area (C.C.A.) is 13209 Ha. The source of water for irrigation is the tail race discharge of Idamalayar Hydro - Electric Project for which a Dam at Ennakkal, has already been completed. This barrage was constructed in 1960-67. The canal system of the Project consists of a main canal, (32.272 km) long on the right bank of the river Periyar. The main canal bifurcates itself into 2 canal systems. The low level canal, 27.25 km long and the link canal (6.73km) long. Out of the total length of 32.278 km of the main canal, 20.629 km has been completed in different reaches. The work is to be arranged in the remaining portion. The total length of link canal is 7.575km.

Banasurasagar Irrigation Project

The project was commenced in 1971 with an estimated cost of Rs.1137.07 lakhs to irrigate an area of 2800 ha. agriculture land for the second and third crops. The project consist of canal system only viz. 2.73 length, two branches with a total length of 13.76 km, 14 no. of distributaries having a total length of 69.04 km. The project report was revised based on 96 schedule of rates and administrative sanction was given to a cost of Rs.37.88 crores. The revised estimate of the project based on 2008 schedule of rate is Rs.127.00 crores. The work of the main canal of length 2.73 km.except for the acqueduct from Ch.0-150 M. and from Ch. 1130 M to 1500 M. is completed.

Chamravattom Regulator–cum–Bridge

A regulator cum bridge at Chamravattom across Bharathapuzha about 6.5 km upstream of river mouth is a multi-purpose project. The targeted irrigated area is 4344 ha. (Gross) and 3170 ha. (Net). The original estimate was Rs.70.00 crores during the year of commencement (1983). The NABARD has approved the project on 31.01.2008 for Rs.106.00 crores with their share of 95.12 crores. The total expenditure incurred for the project upto 31.03.2010 is Rs.39.65 crores. The project is expected to be completed in 2011-12 and the bridge component will be completed by March 2011.

Irrigation Status

As per the assessment of the Directorate of Economics and Statistics the net irrigated area in the state as on March 2010, is 3.86 lakh ha. and the gross area irrigated is 4.54 lakh ha. The net area irrigated has declined from 3.99 lakh ha during 2008-09 to 3.86 lakh ha in 2009-10. Only 16.34 per cent of the net cropped area is irrigated. The percentage of net area irrigated to net area has declined and percentage of gross irrigated area to gross cropped area records a slight increase during the year compared to the last year. During 2009-10 the net irrigated area registered a decline of 10.75 per cent and gross irrigated area by 0.64 percent compared to the previous year. During 2009-10, among the crops, paddy tops among the major crop supported by irrigation. It accounted for about 37 per cent followed by coconut (33%), banana (8%), arecanut (8%) and vegetables (4%).

The source-wise area irrigated as on March 2013 is *given here* . As per the assessment of the Directorate of Economics and Statistics the net irrigated area in the state as on March 2013, is 3.96 lakh ha. and the gross area irrigated is 4.58 lakh ha. There is a decrease of 3.2 percent in the net irrigated area of the state in compared to the previous year of 2011-12. Gross irrigated area also decreased 6.7 percent during the period. Gross irrigated area to Gross Cropped Area in the period is 17.67 percent. During 2012-13, among the crops, coconut tops among the major crop supported by irrigation. It accounted for about 36 percent followed by paddy 32 percent, banana 10 percent, arecanut 8 percent and vegetable 4 percent. *Details* are *given* here. There has been a good progress in irrigated area under vegetable cultivation during the year and also an increase in the area under irrigation for banana cultivation compared to the previous year.

Table. Net Area Irrigated

(in Ha.)

Sl. No.	Source	2004-05	2005-06	2006-07	2007-08	2008-09	2009-10
1	2	3	4	5	6	7	8
1	Government canals	101397	104669	98664	88318	95956	94813
2	Private canals	4729	4965	4300	4324	6318	2656
3	Tanks	43983	45062	42064	41580	39752	40851
4	Wells	108445	110000	114477	131002	133312	125892
5	Other sources	134802	135227	125900	122321	123915	122118
6	Total	393356	399923	385405	387545	399253	386330
7	Area irrigated more than once in a year		918341	-	-	-	
8	Gross irrigated area	455391	464765	475231	455310	458238	454783
9	Net area irrigated to net area Sown (%)	18	19	17.52	18.41	18.86	16.34
10	Gross irrigated area to gross cropped area (%)	15	15	16.29	16.44	16.96	17.04
11	Irrigated area under paddy to total irrigated area	40	38	45	40	37	37

VEGETATION

The varied topographical features, high rainfall and geologic conditions have favoured the formation of different ecosystems from shola forests on the mountain valleys to the mangrove forests along sea coasts and estuaries. The most outstanding feature of the State is the formation of tropical rainforests along the windward side of the Southern Western Ghats, which is lying parallel to the west coast. A small extent of area of the State is along the rain shadow region the Western Ghats, where the vegetation is dominated by dry deciduous forests and scrub jungles. The wet lands are mostly confined to the low land region of the State.Champion and Seth (1968) recognised 26 forest types in Kerala of which the major ones are the west coast tropical evergreen, west coast semi-evergreen, southern moist mixed deciduous, southern dry mixed deciduous, southern montane wet temperate forests, southern subtropical hill forests, southern montane wet temperate grasslands and littoral forests (mangroves). Certain edaphic types recognised are Bamboo brakes, Cane brakes, Reed brakes, Euphorbiaceous scrub jungles, laterite thorn forests and Myristica swampforests. Based on dynamics they recognised secondary forests such as secondary evergreen, secondary moist deciduous, secondary dry deciduous, etc.

SOUTHERN MOIST MIXED DECIDUOUS FORESTS ....

Kerala is a land of intense cultivation. Several exotic species have been introduced as cash crops. They are raised in pure crops or as mixture along with indigenous species. Coconut, arecanut and paddy are the most extensively raised indigenous species. A variety of herbaceous as well as shrubby species are often grown along with coconut and arecanut. Whereas paddy is raised as a pure crop. Tea, coffee and rubber are the extensively raised exotic cash crops in the State. Tapioca, cashew, coco and tobacco are also cultivated to a large to medium scale. The exotic trees introduced to meet the timber, pulp and paper industry are*Acacia auriculiformis, A. mangium, A. mearnsii, A. dealbata, Eucalyptus grandis, E. tereticornis, E. camaldulensis, Grevillea robusta, Paraserianthes falcataria, Swietenia macrophylla*, etc. Exotic ornamental and avenue trees are *Bauhinia purpurea, Callistemon citratus, Cananga odorata, Cassia siamea, C. nodosa, Couropita guianensis, Delonix regia, Jacaranda mimosifolia, Kigelia africana, Kleinhovia hospita, Peltophorum pterocarpum, Plumeria rubra, Polyalthia longifolia, Spathodea campanulata*, etc. The important horticultural plants are *Achras sapota, Averrhoa bilimbi, A. carambola, Diospyros mangostana, Phyllanthus acidus, Psidium guajava, Syzygium aqueum, S. jambos, S. malaccensis*, etc.

A few species which were introduced as garden plants have become naturalised and attained the status of weed. The most widespread and obnoxious weeds are *Ageratina adenophora, Chromolaena odorata, Eichhornia crassipes, Lantana camara, Mikania micrantha, Mimosa diplotricha and Parthenium hysterophorus*. These aggressive weeds are establishing very fast in the terrestrial as well as aquatic ecosystems and are serious threat to the indigenous flora. It is estimated that about 10 per cent of the flora of the State is composed of exotic species.

COASTAL VEGETATION

Kerala has a seacoast of 590 km. The vegetation along the coast is dominated by herbaceous species like*Alternanthera pungens, Alternanthera sessilis, Boerhavia diffusa, Chloris barbata, Cyperus arenarius, Gisekia pharnaceoides, Glinus oppositifolius, Hydrophylax maritima, Hedyotis diffusa, Ipomoea pes-caprae, Launaea sarmentosa, Leucas aspera, Polycarpaea corymbosa, Spermacoce hispida, Spinifex littoreus*, etc.

SOUTHERN HILL-TOP TROPICAL EVERGREEN FORESTS

This forest type is confined to places above 1,500m elevation. There is no stratification of tree canopy into different tiers. The trees are not very large and often highly branched. Epiphytes are common on branches of trees.

The frequently seen trees are *Actinodaphne bourdillonii, Actinodaphne campanulata, Aglaia bourdillonii, Ardisia rhomboidea, Bhesa indica, Elaeocarpus venustus, Eugenia discifera, Garcinia imberti, Garcinia rubro-echinata, Garcinia travancorica, Gordonia obtusa, Isonandra candolleana, Mastixia arborea, Meliosma pinnata ssp. barbulata, Neolitsea fischeri, Neolitsea scrobiculata,*

Poeciloneuron indicum, Syzygium cumini, Syzygium densiflorum, Syzygium rubicundum, Turpinia nepalensis, Vernonia travancorica, etc.

The lower shrubby layer is composed of *Ardisia blatteri, Euonymus paniculatus, Goniothalamus wightii, Octotropis travancorica, Polyscias acuminata, Strobilanthes luridus, Strobilanthes asper, Strobilanthes tristis, Symplocos wynadense*, etc.

MANGROVES OR TIDAL SWAMP FORESTS

Mangroves are mostly confined to the estuaries and banks of backwaters where the influence of tidal waves is pronounced. Kerala had a substantial area under mangroves. However, due to alternate land use, this unique vegetation now occupies only 17 km2 area (Basha, 1991). Common trees in the mangroves are*Avicennia marina, Avicennia officinalis, Barringtonia racemosa, Bruguiera cylindrica, Bruguiera gymnorrhiza, Cerbera manghas, Ceriops decandra, Excoecaria agallocha, Rhizophora apiculata, Rhizophora mucronata, Sonneratia apetala*, etc. Climbers commonly associated are *Dalbergia candenatensis, Derris trifoliata, Ipomoea alba*, etc.

SOUTHERN DRY MIXED DECIDUOUS FORESTS (DRY DECIDUOUS FORESTS)

The forest type is characterised predominantly by hardwood deciduous tree species. The canopy is open with poor undergrowth. Bamboos are barely represented. The canopy level is vague in this type also. The lower storey consists of mostly shrubs and small trees.

The most characteristic species present invariably in the forest type are *Acacia ferruginea, A. leucophloea, Albizia amara, Anogeissus latifolia, Boswellia serrata, Canthium coromandelicum, Cassia fistula, Chloroxylon swietenia, Commiphora pubescens, Diospyros ovalifolia, D. cordifolia, Dodonaea angustifolia, Garuga floribunda, Hardwickia binata, Manilkara hexandra, Premna tomentosa, Santalum album, Sapindus emarginatus, Shorea roxburghii, Sterculia urens, Tarenna asiatica*, etc.

SOUTHERN MOIST MIXED DECIDUOUS FORESTS ....

This forest type is seen below 700 m. During wet season, because of the thick foliage, the canopy looks similar to that of semi-evergreen forests and therefore scarcely distinguishable. However, during dry season the moist deciduous forests reveal their true identity as the trees shed their leaves. The leafless period varies from a few weeks up to 5 months depending on the species. Among the trees, *Bombax insigne, Hymenodictyon obovatum* and *Lagerstroemia microcarpa* have a leafless period up to five months. Terminalia paniculata and Dalbergia sissoides have a leafless period of less than 2 weeks. In Moist deciduous forests also three tier stratification for trees can be met with.

The trees in the upper stratum are *Albizia lebbeck, A. odoratissima, A. procera, Alstonia scholaris, Bombax ceiba, B. insigne, Dalbergia sissoides, Dillenia pentagyna, Gmelina arborea, Grewia tiliifolia, Haldina cordifolia, Hymenodictyon orixense, Lagerstroemia microcarpa, Lannea coromandelica, Melia dubia, Pterocarpus marsupium, Radermachera xylocarpa, Stereospermum colais, Tectona grandis, Terminalia bellirica, T. elliptica, T. paniculata, Tetrameles nudiflora and Xylia xylocarpa*.

The middle stratum is composed mainly of, *Bauhinia malabarica, B. racemosa, Briedelia retusa, Cassia fistula, Careya arborea, Cleistanthus collinus, Dalbergia lanceolaria, Ficus callosa, F. exasperata, F. racemosa, Garuga pinnata, Hymenodictyon obovatum, Macaranga peltata, Miliusa tomentosa, Olea dioica, Phyllanthus emblica, Sapindus trifoliata, Schleichera oleosa, Spondias pinnata, Sterculia guttata, S. villosa, Streblus asper, Strychnos nux-vomica, Trema orientalis, Zanthoxylum rhetsa*, etc.

At the lower stratum, *Casearia tomentosa, Cipadessa baccifera, Cochlospermum religiosum, Grewia glabra, G. nervosa, Helicteres isora, Holarrhena pubescens, Naringi crenulata, Securinega leucopyrus, Tabernaemontana heyneana, Tamilnadia uliginosa, Wrightia tinctoria,* etc. are common.

The climbers include *Acacia pennata, A. torta, Bauhinia scandens var. anguina, Briedelia scandens, Calycopteris floribunda, Cissus heyneana, C. latifolia, Dalbergia volubilis, Olax imbricata, Spatholobus roxburghii, Ziziphus rugosa*, etc.

The undergrowth is composed of *Abutilon persicum, Baliospermum montanum, Barleria prattensis, Chromolaena odorata, Canthium angustifolium, Clerodendrum serratum, Desmodium pulchellum, D. triangulare var. congestum, D. velutinum, Flemingia strobilifera, Lantana camara var. aculeata*, etc.

SOUTHERN MONTANE WET GRASSLANDS

This type is confined to the hill-tops, and the vegetation is dominated by grasses.

The common grasses are *Apocopis courtallica, Arundinella leptochloa, A. mesophylla, Chrysopogon asper, C. hackelii, Cymbopogon flexuosus, Eulalia trispicata, Dimeria ornithopoda, Ischaemum indicum, Jansenella griffithiana, Sacciolepis indica, Themeda triandra, Zenkeria elegans*, etc.

The grasslands also support several herbaceous and subshrubby species. The common subshrubby species are *Hedyotis santapaui, Hypericum mysorense, Impatiens hensloviana, Lobelia nicotianaefolia, Osbeckia aspera, O. leschenaultiana*, *Strobilanthes* spp., etc.

The herbaceous species are *Aeginetia pedunculata, Anaphalis lawii, Andrographis neesiana, Arisaema tortuosum, Cajanus heynei, C. lineata, Cyanotis spp., Dichrocephala integrifolia, Drosera peltata, Dumasia villosa, Emilia scabra, Eriocaulon spp., Crotalaria fysonii, Habenaria spp., Hedyotis swertioides,*

Heracleum candolleanum, Gynura nitida, Impatiens spp., *Lobelia heyneana, Juncus prismatocarpus, Juncus inflexus, Pedicularis perrottetii, Leucas hirta, Malaxis acuminata, Neanotis decipiens, N. indica, N. tubulosa, Peucedanum anamallayense, Pecteilis gigantea, Phyllocephalum courtallense, Plectranthus nilgherricus, Pogostemon rotundatus, Teucrium tomentosum*, etc.

WEST-COAST TROPICAL EVERGREEN FORESTS

This type is characterised by the short-boled and highly branched trees. The branches are densely clothed with moss and other epiphytes. There is no stratification of trees. Leaves of trees in general are small.Climbers are few.

The characteristic species are *Actinodaphne bourdillonii, Cinnamomum sulphuratum, Elaeocarpus munronii, Elaeocarpus recurvatus, Euonymus indicus, Fagraea ceylanica, Gordonia obtusa, Ligustrum robustum ssp. walkeri, Maesa indica, Mahonia leschenaultii, Michelia champaca, Pittosporum neelgherrense, Rapanea thwaitesii, Rhododendron arboreum ssp. nilagiricum, Symplocos cochinchinensis ssp. laurina, Syzygium densiflorum, Turpinia nepalensis*, etc.

The common epiphytes are *Aerides ringens, Bulbophyllum fischeri, Coelogyne nervosa, Eria reticosa, Impatiens auriculata, Impatiens parasitica, Liparis elliptica, Oberonia santapaui, Trichoglottis tenera*, etc.

SOUTHERN TROPICAL THORN FORESTS (SCRUB JUNGLES)

This forest type is restricted to the rain shadow regions in Idukki and Palakkad districts. The species are mostly xerophytic with short bole and low branching. The canopy is wide open and the canopy level differentiation is not distinct. The hardwood trees, thorny shrubs and climbers are characteristic features of the forest type. The undergrowth is furnished with some herbaceous forms during monsoons and remains exposed for the rest of the time.The major species representing the forest type are *Acacia spp., Albizia amara, Atalantia monophylla, Capparis spp., Caralluma spp., Carissa carandas, Ceropegia juncea, Cordia spp., Dichrostachys cinerea, Diospyros cordifolia, Euphorbia spp., Grewia spp., Helixanthera spp., Opuntia spp., Pergularia daemia, Pisonia aculeata, Pleiospermium alatum, Prosopis juliflora, Strychnos potatorum, Ziziphus spp.*, etc.

WETLANDS AND AQUATIC VEGETATION

These vegetation types are confined to lowlands and midlands. In the ponds, lakes, lagoons and marshy areas where plants like *Aponogeton natans, Ceratophyllum demersum, Hydrilla verticillata, Nelumbo nucifera, Nymphaea nouchali, Nymphaea pubescens, Nymphoides cristata, Nymphoides macrosperma, Ottelia alsinoides, Utricularia spp., thrive. While species like Najas graminea, Polypleurum spp., Zeylanidium olivaceum, etc. are attached to submerged rocks in the rivers. Semiaquatic species such as Eriocaulon spp., Hydrophylla schulli,*

Hydrolea zeylanica, Monochoria vaginalis, Lindernia spp., etc. are found in marshy areas and banks of ponds and lake shores.

WEST-COAST SEMI-EVERGREEN FORESTS

This forest type is found below 700 m asl. It is intermediate between evergreen and moist deciduous types where several species which are common in evergreen forests and some of the trees in the moist deciduous forests are also seen. In this type also a three layer stratification of the trees are met with.

The upper stratum is composed mainly of *Aglaia barberi, Antiaris toxicaria, Artocarpus hirsutus, A. gomezianus ssp. zeylanicus, Bischofia javanica, Bombax ceiba, Carallia brachiata, Chukrasia tabularis, Dimocarpus longan, Diospyros buxifolia, D. crumenata, Drypetes confertiflora, Dysoxylum beddomei, Hopea parviflora, Mimusops elengi, Polyalthia fragrans, Prunus ceylanica, Pterospermum reticulatum, P. rubiginosum, Sageraea dalzellii, Otonephelium stipulaceum, Terminalia bellirica, Tetrameles nudiflora, Toona ciliata, Vitex altissima*, etc.

At the middle stratum, *Aglaia barberi, Aglaia lawii, Aporusa lindleyana, Diospyros bourdillonii, D. assimilis, D. montana, D. paniculata, Flacourtia montana, Ficus callosa, F. nervosa, Harpullia arborea, Holigarna grahamii, Holoptelea integrifolia, Hydnocarpus alpina, H. pentandra, Margaritaria indica, Nothopegia colebrookeana, Olea dioica, Oroxylum indicum, Pajanelia longifolia, Schleichera oleosa*, etc.

The lower stratum trees are *Acronychia pedunculata, Agrostistachys indica, Antidesma menasu, Callicarpa tomentosa, Drypetes oblongifolia, Hunteria zeylanica, Ixora brachiata, Mallotus philippensis, Meiogyne ramarowii, Memecylon spp., Sapindus trifoliata, Xanthophyllum arnottianum*, etc.

The woody climbers include *Acacia concinna, A. pennata, Ancistrocladus heyneanus, Anodendron manubriatum, Bauhinia scandens var. anguina, Caesalpinia cucullata, Chilocarpus denudatus, Erycibe paniculata, Olax imbricata, Salacia oblonga, Sarcostigma kleinii, Strychnos colubrina, S. minor*, etc.

The lower shrubby layer is composed of *Barleria courtallica, Dichapetalum gelonioides, Dracaena terniflora, Gomphandra tetrandra, Glycosmis* spp., *Justicia santapaui, Leea indica, Murraya paniculata, Strobilanthes* spp., etc.

WEST-COAST TROPICAL EVERGREEN FORESTS..........

Evergreen forests are characterised by the profusion of species, particularly trees and woody climbers and the canopy is closed. The secondary species and exotic weeds are absent. The lower shrubby layer is composed mostly of seedlings and saplings of tree species. The evergreen forests are found between 400 - 1,200 m altitude. It has been observed that there is difference in the physiognomy and composition of species with respect to altitude. The evergreen forests below 700 m are occurring in patches and mostly confined to the sides of water courses where the soil is rather deep. Though there is no marked dominance of any group of trees in evergreen forests below 700 m elevation, there is dominance of certain species in the upper stratum above 700 m forming associations. The important associations are: (i) *Vateria - Calophyllum - Cullenia,* (ii) *Mesua - Cullenia - Palaquium* and (iii) *Mesua - Palaquium*. In the forests above 1,100 m there is dominance of Lauraceae and *Heritiera papilio*. Epiphytes and mosses tend to increase with altitude while woody climbers decrease.

Species composition below 700 m altitude

The upper stratum trees are *Antiaris toxicaria, Artocarpus heterophyllus, Bischofia javanica, Canarium strictum, Chrysophyllum roxburghii, Cynometra travancorica, Diospyros buxifolia, Dipterocarpus indicus, Drypetes elata, Dysoxylum beddomei, Elaeocarpus tuberculatus, Hopea parviflora, Kingiodendron pinnatum, Mangifera indica, Mastixia arborea ssp. meziana, Ormosia travancorica, Persea macrantha, Syzygium chavaran, S. gardneri, Terminalia travancorensis, Vateria indica*, etc.

The middle stratum trees are *Aglaia lawii, A. malabarica, A. perviridis, Alseodaphne semecarpi-folia var. parvifolia, Aphanamixis polystachya, Beilschmiedia bourdillonii, Cinnamomum mala-batrum, Diospyros spp., Elaeocarpus glandulosus, E. serratus, Garcinia gummi-gutta, G. morella, G. spicata, Gymnacranthera farquhariana, Holigarna arnottiana, Hydnocarpus alpina, Knema attenuata, Lepisanthes tetraphylla, Madhuca neriifolia, Myristica beddomei, M. malabarica, Otonephelium stipulaceum, Reinwardtiodendron anamalaiense, Sageraea dalzellii, Strombosia ceylanica, Syzygium laetum, Vepris bilocularis, Walsura trifolia*, etc.

The lower stratum trees are *Agrostistachys indica, Atalantia racemosa, Baccaurea courtallensis, Casearia ovata, Isonandra lanceolata, Leptonychia caudata, Meiogyne ramarowii, Nothopegia racemosa, Orophea erythrocarpa, O. uniflora, Psydrax dicoccos, P. travancorica, Turraea villosa* and pole crops of upper and middle stratum trees.

The woody climbers are *Acacia concinna, Aganope thyrsiflora, Artabotrys zeylanicus, Bauhinia phoenicea, Caesalpinia cucullata, Combretum latifolium, Derris brevipes, Desmos lawii, Entada rheedei, Sarcostigma kleinii, Spatholobus purpureus, Strychnos lenticellata, Tetracera akara, Ventilago bombaiensis*, etc.

The main undergrowth are *Alpinia malaccensis, Barleria courtallica, Blachia denudata, Ixora nigricans, Munronia pinnata, Psychotria* spp., *Strobilanthes* spp., etc.

Species Composition Above 700 M altitude

The upper stratum trees are *Artocarpus heterophyllus, Calophyllum polyanthum, Chryso-phyllum roxburghii, Cullenia exarillata, Dysoxylum malabaricum, Elaeocarpus tuberculatus, Fahrenheitia zeylanica, Holigarna ferruginea, Mesua ferrea, Palaquium ellipticum, Persea macrantha, Poeciloneuron indicum, Prunus ceylanica, Toona ciliata, Vateria indica*, etc.

The middle stratum trees include *Actinodaphne malabarica, Agrostistachys borneensis, Beilschmiedia wightii, Bhesa indica, Dimorpho-calyx lawianus, Diospyros paniculata, D. sylvatica, Drypetes malabarica, Melicope lunu-ankenda, Glochidion ellipticum, Gordonia obtusa, Hydnocarpus macrocarpus, Litsea bourdillonii, L. floribunda, L. oleoides, Mallotus tetracoccus, Meliosma pinnata, M. simplicifolia, Trichilia connaroides, Turpinia malabarica*, etc.

Aglaia tomentosa, Aporusa acuminata, Ardisia pauciflora, Casearia ovata, Dendrocnide sinuata, Gomphandra coriacea, Lepisanthes erectus, Ligustrum robustum ssp. walkeri, Meiogyne pannosa, Oreocnide integrifolia, Orophea uniflora, Psychotria anamallayana, Sarcococca coriacea, Symplocos cochinchinensis ssp. laurina, Symplocos macrophylla spp. rosea, Syzygium munronii, etc. are common in the lower stratum.

The climbers are *Aganosma cymosa, Calamus gamblei, C. pseudotenuis, C. vattayila, Carissa spinarum, Embelia ribes, Kunstleria keralense, Millettia rubiginosa, Pseudaidia speciosa, Rourea minor, Sabia malabarica, Strychnos lenticellata, Toddalia asiatica*, etc.

The undergrowth is mainly composed of *Amomum canniaecarpum, A. muricatum, Aralia malabarica, Elettaria cardamomum, Lasianthus rostratus, Lepianthes umbellata, Munronia pinnata, Mycetia acuminata, Pandanus thwaitesii, Pavetta calophylla, Psychotria flavida, P. nigra, P. nudiflora, Rhynchotechum permolle, Tarenna monosperma*, etc.

In the forests above 1,200 m the trees are pre-dominantly of Lauraceae. The other trees are *Calophyllum austroindicum, Heritiera papilio, Mesua thwaitesii*, etc.

THE WESTERN GHATS BY MOHAN PAI - TOPOGRAPHY

The Western Ghats form an almost unbroken rampart on the fringe of western peninsula parallel to the west coast for about 1600 km and often hardly 40 km from the Arabian Sea.

They start immediately south of the Tapti river, the northern most point being the Kundaibari pass (21006N, 74011E) near Brahmavel in Dhule district of Maharashtra ending near Kanya-kumari (80N) barely 20 km from the sea in Tamil Nadu. They cover an area of approximately 1,59,000 sq. km with an average elevation of 900-1500 m. ASL, obstructing the monsoon winds from the south west and the orographic effect is considerable. Although the average heights of the Ghats is less than1500 m. ASL, in the southern reaches it rises 2000 m and to exceptionally higher peaks of 2,500 m and above.

Along its entire length, the Western Ghats range has only one total discontinuity, the Palghat Gap in Kerala where for more than 30 km there is a gap which has a floor height of less than 100 m ASL. This discontinuity is perhaps of tectonic origin through which a river may have flowed in ancient times. The peninsular plateau is highest in the south and west and slopes eastward, the eastern edge forming the broken up Eastern Ghats. The Eastern and the Western Ghats meet along the Moyar Gorge with the Billigirirangana Hills along the north-eastern side and the the Nilgiris in the south-west.

Based on the topography and geology, the Western Ghat region is divided into three distinct subregions:

NORTHERN WESTERN GHATS (TAPTI RIVER TO GOA)

This region consists of the most homogeneous part of the Western Ghats, hugging the coast for almost 600 km. It corresponds to the western edge of the vast plateau formed by massive horizontal outflow of volcanic lava which cooled to form dark grey basalt.

Fig. *The layer of lava often interlaced with non-volcanic debris.*

The scarp of the Ghats in this region presents a magnificent profile of over 1,000 m of successive volcanic layers, which on erosion have produced a typical trappean landscape, forming a formidable well-dissected wall looking over the narrow west coast plains, buton the eastern side descending in steps one below the other. It is in this region that the full significance of the term ghats (steps of a stair-case) becomes clear. The layers of lava are quite often interlaced with non-volcanic debris. Sometimes these form intertrappean deposits holding plant and animal fossils. The elevation is generally between 700 and 1,000 m, but some of the pinnacles attain greater heights; the tallest are the Kalsubai (1,646 m) near Igatpuri, Salher (1,567 m) 90 km north of Nasik and the famous Mahabaleshwar (1,438 m).

The escarpment is not a simple erosional feature; some geologists believe that it marks the location of a broad zone of en echelon deep-seated faults. Landsat imagery shows a large density of faults striking NW-SW along the trend of this escarpment roughly parallel to the coast, in which 33 hot springs have been noted and which have been interpreted as indicative of a fault. The coastal zone here called Konkan, is a narrow strip about 50-60 km wide. It is made up of a series of more or less high hills, some of them like Matheran (700 m) almost reaching the height of the plateau and bears testimony of regressive erosion.

Central Western Ghats (Goa to Nilgiris)

The basaltic outpourings cease to the north of Goa. The Middle Western Ghats run from a little south of 160N latitude up to Nilgiri Hills. Towards the south, the Ghats consist of complex formation of pre-cambrian rocks. In the central Western Ghats, the rocks are predominantly of Dharwar system (among the oldest in India) and Peninsular gneisses.

The Western scarp is considerably dissected by headward erosion of the west flowing streams. The elevation generally range between 600 to 1000 m up to 13030N. The Ghats lose their graded appearance and form a steep barrier whose height becomes moreirregular. They rise suddenly at Kodachadri (1343 m)and fall to about 600 m at Agumbe.

Fig. *The Iron mountains- Kudremukh range*

From Kudremukh (1,892 m) up to Palghat Gap, the edge of the plateau is very often higher than 1,000 m. and the peaks become more numerous and higher - Pushpagiri (1,713 m) in the North Kodagu, Tadianamol Betta (1,745 m), Banasuram (2,060 m), Vavul Mala (2,339 m) at the edge of the Wayanad plateau. Towards 11030N, the Western Ghats composed of hard Charnokites, rise abruptly in the Nilgiri horst where they join the Eastern Ghats. The Nilgiri mountains constitute an elevated plateau dominated by two of its highest peaks, the Dodabetta (2,637 m) and Makurti (2,554 m) overlooking Palghat Gap from a height of more than 2,000 m.

On the Mysore plateau, whose average elevation range from 700 to 900 m we find reliefs formed by tectonic events such as spectacular horseshoe of the Bababudan hills which extends from Hebbe through Kemmanagundi and Attigudi to Mulainagiri (1,923 m) which is the highest peak in Karnataka.

Fig. *Brahmagiris near Iruppu, Kodagu*

The other tallest peak is Bababudangiri (Chandradrona Parvata 1,894 m). The width of the coastal zone is also more variable here than in Maharashtra. It is about 40 km wide at the latitude of Goa and then suddenly narrows near Karwar where the Ghats dip into the seawith peaks emerging as picturesque islands.

Fig. *The Ghats dip into the sea near Karwar*

This advance of the relief is carved by deep valleys of the Kalinadi, Gangavali and Sharavathy. The last drops from a height of 250 m creating the famous Jog falls.

To the south of 140 N, the coastal zone now called South Kanara, widens once more to about 80 km. The coastal region after Kodagu known as Malabar is not more than 30 km wide up to the latitude of Kozikode. From here it widens out to about 60 km till the Palghat Gap. The coastal hills in the entire region, particularly to the north of Mangalore are mostly tabular relief hardened by iron oxides. These reliefs are practically bare and present a characteristic landscape.

Fig. *Charmadi Ghats*

Southern Western Ghats (South of the Palghat Gap)

The Western Ghats are separated from the main Sahyadri Range by the Palghat Gap which is about 30 km wide and they appear abruptly as the Anaimalai-Palni block whose high plateau attain a height of 2,695 m in the Anaimudi peak, the highest point in south India.

This block is a composite range ismade up of the Nelliampathy plateau (drained by Chalakudi) to the west, the Anaimalai plateau (largely converted into tea plantations and distinctly elevated to the east) in the centre and the Palni horst overlooking the peneplain of Tamil Nadu from a height of almost 2,000 m.

To the south of this west-east oriented block, the Ghats display further changes. Here they form an elevated plateau slanting towards the west - the Periyar plateau, thus named after its most important river. The eastern part of this plateau forms Elamalai range, better known as Cardamom Hills because of its plantations. This central range attains its peak at Devar Malai (1,922 m) and terminates in the east by sheer cliff 1,000 m high. From this, the SW-NE oriented Varushanad massif is detached and continued by the Andipathi, which together with Palni hills embraces the Kambam Valley.

South of Devar Malai, at about 90N, the Ghats are once again interrupted by narrow Shencottah Pass (alt. 160 m). From here they continue as a narrow ridge with steep slopes to the west as well as to the east, until about 20 km before Kanyakumari. This last bit is very rugged and its highest peak is the Agasthyamalai (1,869 m). Three regions may be distinguished here; Agasthyamalai proper, Mahendragiri to the south and the Tirunelveli hills on the eastern slopes. The coastal zone (30-50 km wide) constituting Travancore is made up of convex shape hills with rounded summits. Here we do not find the tabular reliefs of South Kanara and Konkan since their formation at this latitude is probably more difficult due to short, dry season.

The vegetation types are characterised by low level tropical evergreens turn to Shola grasslands on the cool wind swept slopes of the higher ranges. Primary or secondary moist deciduous forests cover the lower western hills. Moist deciduous forests are also common on the eastern slopes in the rain shadow area. Dry deciduous and even scrub vegetation characterises the eastern slopes where the humidity is low and the winds are high.

The Imperial Gazetteer of India - 1907 gives a very vivid description of the region which is reproduced below:

Western Ghats - a range of mountains about 1,000 miles(1,600 km) in length, forming the western boundary of the Deccan and the watershed between rivers of peninsular India. The Sanskrit name is Sahyadri.

The range, which will be treated here with reference to its course through Bombay, Mysore and Coorg and Madras, may be said to begin at the Kundaibari pass in the south western corner of the Khandesh district of Bombay Presidency, though the hills that run eastward from the pass to Chimtana and overlook the lower Tapti valley, belong to the same system. From Kundaibari (21006N, 740 11E) the chain runs southward with an average elevation which seldom exceeds 4,000 ft., in a line roughly parallel with the coast, from which its distance varies from 20 to 65 miles. For about 100 miles up to a point near Trimbak, its direction is somewhat west of south; and it is flanked on the west by the thickly wooded and unhealthy tableland of Peint, Mokhada and Jawhar (1500 ft) which forms a steep barrier between the Konkan lowlands and the plateau of the Deccan (about 2000 ft). South of Trimbak the scarp of the western face is more abrupt; and for 40 miles, as far as the Malsej pass, the trend is south-by-east changing to south-by-west from Malsej to Khandala and Vagjai (60 miles), and again to south by east from here until the chain passes out of the Bombay Presidency into Mysore near Gersoppa (14010N, 74050E).

On the eastern side the Ghats throw out many spurs or lateral ranges that run from west to east, and divide from one another the valleys of the Godavari, Bhima and Kistna river systems. The chief of these cross ranges are Satmalas, between the Tapti and Godavari valleys; the two ranges that break off from the main chain near Harishchandragarh and run south eastwards into the Nizams Dominions, enclosing the triangular plateau on which Ahamad-nagar stands, and which is the watershed between the Godavari and the Bhima; and the Mahadeo range, that runs eastward and southward from Kamalgarh and passes into the barren uplands of Atpadi and Jath, forming the watershed between the Bhima and the Kistna systems. North of the latitude of Goa the Bombay part of therange consists of Eocene trap and basalt, often capped with laterite, while farther south are found such older rocks as gneiss and transitional sandstones. The flat-topped hills, often crowned with bare wall like masses of basalt, or laterite are clothed on their lower slopes with jungles of teak and bamboo in the north; with jambul (eugenia jambolana), ain (Terminalia tomentosa) and nana (Lagerstroemia parviflora) in the centre; and with teak, blackwood, and bamboo in the south.

On the main range and its spurs stand a hundred forts, many of which are famous in Maratha history. From north to south, the most notable points in the range are the Kundaibari pass a very ancient trade route between Broach and the Deccan; the twin forts of Salher and Mulher guarding the Babhulna pass; Trimbak at the source of holy river Godavari; the Thal pass by which the Bombay-Agra road and the northern branch of the Great Indian Peninsula Railway ascends the Ghats; the Pimpri pass, a very old trade route south between Nasik and Kalyan or Sopara, guarded by the twin forts of Alang and Kulang; Kalsubai (5427 ft), the highest peak in the range; Harishchandragarh (4691 ft); the Nana pass, a very old route between Junnar and Konkan; Shivner, the fort of Junnar; Bhimashankar, at the source of the Bhima; Chakan, an old Musalman stronghold; the Bhor or Khandala pass, by which the Bombay-Poona road and the southern branch of the Great Indian Peninsula Railway enters the Deccan, and on or near which are the caves of Kondane, Karli, Bhadja and Bedsa; the caves of Nasdur and Karsambla below the forts of Sinhagarh and Purandhar in the spurs south of Poona; the forts of Raigarh in the Konkan and of Pratapgarh between the new Fitzgerald ghat road and the old Par pass; the hill station of Mahabaleshwar (4717 ft) at the source of the Kistna; the fort and town of Satara; the Kumbharli pass leading to the old towns of Patan and Karad; the Amba pass, through which runs the road from Ratnagiri to Kolhapur; the forts of Vishalgarh and Panhala; the Phonda pass, through which runs the road from Deogarh to Nipani; the Amboli and the Ram pass, through which run two made roads from Vengurla to Belgaum; Castle Rock, below which passes the Railway from Marmagao to Dharwar; The Arabail pass on the road from Karwar to Dharwar; the Devimane pass on the road from Kumta to Hubli, and the Gersoppa Falls on the river Sharavati.

On leaving the Bombay Presidency, the Western Ghats bound the State of Mysore on the west, separating it from the Madras district of South Kanara, and run from Chandragutti (2,794 ft) in the north-west to Pushpagiri on the Subramanya hill (5,626 ft) in the north Coorg and continue through Coorg into Madras.

In the west of the Sagar taluk, from Govardhangiri to Devakonda, they approach within ten miles of the coast. From there they trend south-eastwards, culminating in Kudremukh (6,215 ft) in the southwest of Kadur district, which marks the watershed between Kistna and Cauvery systems. They then bend east and south to Coorg, receding to 45 miles from the sea. Here to numerous chains and groups of lofty hills branch off from the Ghats eastwards, forming the complex series of mountain heights south of Nagar in the west of Kadur district. Gneiss and hornblende schists are the prevailing rocks in this chapter, capped in many places by laterite, with some bosses of granite. The summits of the hills are mostly bare, but the sides are clothed with magnificent evergreen forests. Ghat roads to the coast have been made through the following passes: Gersoppa, Kollur, Hosangadi, and Agumbe in Shimoga district; Bundh in Kadur district, Manjarabad and Bisale in Hassan district.

In the Madras Presidency, the Western Ghats continue in the same general direction, running southwards at a distance of from 50 to 100 miles from the sea until they terminate at Cape Comorin, the southern most extremity of India. Soon after emerging from Coorg they are joined by the range of the Eastern Ghats, which sweeps down from the other side of the peninsula; and at the point of junction they rise up into the high plateau of the Nilgiris, on which stand the hill stations of Ootacamund (7,000 ft), the summer capital of the Madras Government, Coonoor, Wellington, and Kotagiri and whose loftiest peaks are Dodabetta (8,760 ft) and Makurti (over 8,000 ft).

Immediately south of this plateau the range, which now runs between the districts of Malabar and Coimbatore, is interrupted by the remarkable Palghat Gap, the only break in the whole of its length. This is about 16 miles wide, and is scarcely more than 1,000 ft above the level of the sea. The Madras Railway runs through it, and it thus forms the chief line of communication between the two sides of this part of the peninsula.

South of this gap the Ghats rise abruptly again to even more than their former height. At this point they are known by the local name Anaimalais, or elephant hills, and the minor ranges they here throw off to the west and east are called respectively the Nelliam-pathis and the Palni Hills. On the latter is situated the sanatorium of Kodaikanal. Thereafter, as they run down to Cape Comorin between the Madras Presidency and the native state of Travancore, they resume their former name.

North of the Nilgiri plateau the eastern flank of the range merges somewhat gradually into the high plateau of Mysore but its western slopes rise suddenly and boldly from the low coast south of the Palghat Gap both the eastern and western slopes are steep and rugged . The range here consists throughout of gneisses of various kinds, flanked in Malabar by picturesque terraces of laterite which shelve gradually down towards the coast. In elevation it varies from 3,000 to 8,000 ft above the sea, and the Anaimudi peak (8,839 ft) in Travancore is the highest point in the range and in southern India. The scenery of the Western Ghats is always picturesque and frequently magnificent, the heavy evergreen forest with which the slopes are often covered aiding greatly to their beauty. Large games of all sorts abounds, from elephants, bisons and tigers to the Nilgiri ibex, which is found nowhere else in India.

Before the days of roads and railways the Ghats rendered communication between the west and east coasts of the Madras Presidency a matter of great difficulty; and the result has been that the people of the strip land which lies between them and the sea differ widely in appearance, language, customs, and laws of inheritance from those in the eastern part of the Presidency. On the range itself, moreover, are found several primitive tribes, among whom may be mentioned the well known Todas of the Nilgiris, the Kurumbas of the same plateau, and the Kadars of Anaimalais. Communications across this part of the range have, however, been greatly improved of late years. Besides the Madras Railway already referred to, the line from Tinnevelly to Quilon now links up the two opposite shores of the peninsula, and the range is also traversed by numerous ghat roads. The most important of these latter are the Charmadi ghat from Mangalore in South Kanara to Mudgiri in Mysore; The Sampaji ghat between Mangalore and Mercara, the capital of Coorg; the roads from Cannanore and Tellichery, which lead to the Mysore plateau through the Perumbadi and Peria passes; and the two routes from Calicut to the Niligiri plateau up the Karkur and Vayittiri-Gudalur ghats.

5

Flowers and Plants of Thrissur and Kerala

Kerala obtain good rain and has a tropical climate which is very suitable for growing flowering plants. Due to the climate in Kerala, there are many types of flowers found abundantly and naturally. Some flowers are common whereas there are some which are rare. There are flowers grown in home gardens and some are found only in the forest areas. Different flowers are supported during different climates. The flowers are used for preparing various ayurvedic medicines as they have many medicinal properties. Flowers are also used while performing various prayers in temples and churches. There are some flowers that are poisonous and have pungent smell.

Kerala's climate is so blessed that many types of flowers grow abundantly and naturally in the soil. Many flowers are so common and some are very rare. Some flowers are found on almost all household, while some are found only in specific locations of forests. There are different types of orchids that can't be found elsewhere outside the silent valley forest of Kerala. Different climates support different types of plants and flowers. Almost all types of flowers in south India are found in Kerala too. Trekkers identify new variety of orchids from the forests of Kerala. Flowers have a prominent place in the pooja rooms (prayer rooms) of Kerala, temples and churches. The most common types of flowers are china rose (hibiscus), rose, orchids and a wide variety of wild flowers. Many of the flowers are of interest to Ayurveda practitioners because of their medicinal properties. The most common hibiscus flowers form a part of natural herbal shampoo that cleans the heads and hair of village women. Some flowers are poisonous too. Such flowers usually have a repelling smell too. The flowers of many plants go into the production of Ayurvedic medicines, along with leaves, seeds, fruits and roots.

AAMBAL

It is also called as water lily or star-lotus. It is an aquatic plant having bright coloured flowers. The stem is submerged in ground and flat round leaves are seen above the water level. The flowers have many petals and floats over the water surface.

Ambal (Nymphaea stellata), also known as water lily or blue lotus is a common flowering water flower, as much popular as lotus in India. According to Indian mystics, lotus is the better half of sun while water lily is of moon. The aquatic plant has roots firmly on the bottom of a pond or river, while the leaves and flowers float above water-level. There are different varieties of water lilies found in Indian water bodies. The plant is also found as an ornamental plant in almost all over the world. Aambal is the Malayalam word for the plant. Water lily is used in different Ayurvedic medicine preparations too. The rhizome is the main usable part. It is used in the treatment of diarrhea, certain skin diseases, piles, and diseases that affect urinary tract and kidney. The leaves and seeds of the plant are used in stomach upset.

ARALI

It is also called by many names such as Nerium plant, Dog-bane, South-sea rose, Adalpha etc. It is a shrub grown in garden and has beautiful flowers. They are also seen as herbs, shrubs, trees and climbers. A unique feature of the members of this family is the production of profuse sap. Even though this sap is poisonous, they are of much medicinal value.

Nerium Oleander (*called as Arali in Tamil*) is a perennial ornamental plant that is highly drought resistant. If you live in a moderate to extremely hot place with poor soil, Nerium Oleander should be the perfect start. They are flexible to the soil and climate conditions and bless the garden with sweet smelling showy flowers.

These plants have been planted through most of the National Highways in Tmail nadu, India. Its actually a bliss to see them flowering all through the dividers in the roads.

ARUMASAM

It is also known as Krishna kireedam or Pagoda plant and is a shrub grown in the wild. Their terminal inflorescence last for about 6 months and the orange-red- bell shaped flowers get opened row by row. It is called as 'Krishna kireedam' which means the crown of lord Krishna because the inflorescence is shaped in such a way.

ASHOKAM

It is also known as Galasoka, Anganapriya or Ashoka tree. It is considered to be a holy tree and its leaves and flowers are used for preparing various ayurvedic medicines. It flowers during all the seasons and is found usually in the courtyard of houses. They provide shade during hot seasons for the people and so it is called by the name 'Angana priya' which means darling of the front yard.

BOTANICAL NAME: SARACA INDICA OR JONESIA ASHOK.

Ashoka is one of the most legendary and sacred trees of India, and one of the most fascinating flowers in the Indian range of flower essences. It belongs to Caesalpaeniaceae family. It is a very handsome, small, erect evergreen tree, with deep green foliage and very fragrant, bright orange-yellow flowers, which later turn red. The flowering season is around April and May. It is found in central and eastern Himalayas as well as on the west coast of Bombay.

Ashoka is a Sanskrit word meaning *without grief or that which gives no grief.* Of course, the tree has many other names in local languages as well. One such name means the tree of love blossoms. The Hindus regard it as sacred, being dedicated to Kama Deva, God of Love. The tree is a symbol of love. Its beautiful, delicately perfumed flowers are used in temple decoration. There are also festivals associated with this flower. Lord Buddha was born under the Ashoka tree, so it is planted in Buddhist monasteries. In India, drinking the water in which the flowers have been washed is widely considered a protection against grief. In a certain sense, such water is a flower essence, its effects recognized by our people for centuries. As for the essence, indeed it is for those who have gone through great trauma and suffering.

It is a healer for deep seated sorrow, sadness, grief, and disharmony in one's inner being due to events such as bereavement, failure, suffering, disease, and isolation. On using this essence, a profound inner state of joy, harmony and well being is produced. It works very gently, in that it changes one's perception of the sorrow.

We have found Ashoka to be especially helpful to the elderly. One woman from New York who was around 76 had gone through much suffering and trauma in life. At this stage, she was all by herself, very exhausted, and worried about her future. She took Ashoka and sounded very positive the next time she called us. Subsequently, she made some positive decisions and made many changes in her life.

Ashoka is also seen as a remedy for women, allowing them to be feminine. The tree is regarded as a guardian of female chastity. The *Vrikshadevatas*—the gods of trees who represent fertility—are known to dance around the tree, and are worshipped by childless women.

Herbally, the bark of this tree is a household remedy for uterine disorders. The essence also helps women to be fertile. It is said that 'weeping woman, weeping womb,' in that the woman's emotional state affects her reproductive organs. Therefore, the essence, like the herb, helps in the uterine problems like excessive bleeding, irregular menstrual periods and infertility.

An example of someone who was much helped by Ashoka was a young woman monk, who would bleed heavily during her menstrual periods. In fact, if she was upset about something, the bleeding would start at once, even if it was not the time for her regular menses. Her nature was very sensitive, and she was easily affected by any small upset. Because of her suffering, she was very irritable and had thoughts of suicide.

Being a monk at a very young age had burdened her with great responsibilities of taking care of others, learning about religion and teaching also. It was a battle for her to go on. At this stage she was also having high blood pressure and dizziness.

The first essence that helped her was Ashoka flower, to overcome the deep-seated grief and sorrow. Her attacks of dizziness and her suicidal thoughts became much less. Her uterus responded as well, with no more excessive bleeding. The success of this treatment motivated her to take further remedies.

ASHOKA TREE BENEFITS

Bark of the tree is used as astringent, demulcent, refrigerant, styptic and febrifuge. Flower of the tree is primarily used as uterine tonic and diabetes for keeping blood sugar under control. Leaves are medicinally used as depurative.

Ashoka tree is a small evergreen herb found in Deccan plateau and Western Ghats of Indian sub-continent. It has beautiful and fragrant flowers which grows in bunches. They are orange and yellow in color which turns into red before they fall. Ashoka gets flowers between February and April. The tree is considered as one of the most sacred trees in the Indian customs. Ashoka tree is dedicated to love god Kama.

The botanical name given to the tree is Saraca Indica and the Indian name is Ashoka tree. The word 'Ashoka' in Sanskrit means 'no grief'. No doubt the meaning is proved by the several medicinal qualities of the tree. Furthermore one can find capsules and tonics made from bark, seeds and flowers of the Ashoka tree. These are sold in market as natural supplements for various disorders. This is the main herb recommended for women to relieve them from gynecological problems. It keeps women energetic and youthful with its anti-depressant properties. Hence it is used as the best medicine for women since ages. In short Ashoka tree is widely used to heal several health issues. Read on the article to know more in detail about the benefits of Ashoka tree.

Health Benefits of Ashoka Tree

Below are some of the top Ashoka tree benefits –

Ashoka tree is a famous medicinal plant which cures ample number of disorders. Each part of the tree is loaded with one or the other health benefits. Listed below are few effective health benefits of Ashoka tree.

- Cosmetic: The bark of the tree is used for preparing cosmetics to improve the complexion. In addition it is also beneficial for ladies who face painful or irregular menstrual cycle. It also best recommended remedy to treat vaginal discharge
- Stomach Swelling: Intake of bark or leaves of Ashoka helps to get rid of worms and removes swelling in the stomach. Furthermore the stem bark is rich in anti-fungal, pain killing and anti-bacterial properties. Hence it provides a great relief from the burning sensation such as scorpion sting
- Blood Purifier: Consuming tonics made of bark, flowers and leaves of Ashoka tree treats diarrhea and purifies the blood

- Diabetes: Dried flowers of the tree are very helpful for the people who are suffering from diabetes. Prepare powder from dry flower of the tree and take 1 tsp daily to keep sugar level in control
- Piles: In addition the Ashoka herbal supplements cure the bleeding caused by piles. You can also prepare a decoction using the bark of the tree to treat internal piles
- Inflammation: Decoction of the bark relieves from burning sensation and can be used as a wash. It also soothes inflammation externally and internally.
- Uterus: The bark of this herbal tree is rich in ketosterol. It treats uterine fibroids and other internal fibroids as well. Hence it is considered as one of the common household remedy for uterine disorders. The mucosa of the uterus could be toned by powdered bark. Boil 2 tsp of bark powder in 2 glass of water till it is reduced to 1/4th. Filter it and take 30ml of this decoction 2 times a day
- Fungal Infection: Ashoka tree bark is loaded with chloroform, methanol and other properties that treat bacterial and fungal infections
- Complexion: Capsules produced from Ashoka tree are of great benefit for skin and overall complexion. In addition these tablets are consumed as natural supplement to treat irritations or burning sensations in the skin

Note and Precautions for Using Ashoka Tree for Health Benefits

There are lots of Ashoka tree benefits and no serious side effects caused with the use of Ashoka supplements, but still one should always keep the following things in mind.

- It is advised that pregnant and breastfeeding mothers should not consume any products produced from Ashoka tree
- It is always safe to consult your health professional before trying out any of these natural supplements

CHEMPAKAM

It is also called as Joy-perfume tree. They are seen as both shrubs and trees. The flowers are shaped like a cup and are fleshy too. The petals and the sepals are not distinguished. The flowers have a very good fragrance and so are used as a room freshener and also for making perfumes. Women here like to wear this flower in their hair due to their high fragrance.

JOY PERFUME TREE

Joy perfume tree (*Magnolia champaca*), formerly *Michelia champaca*, also called champak, champac, orchampaca, tree native to tropical Asia that is best known for its pleasant fragrance. The species, which is classified in the magnolia family (Magnoliaceae), is also characterized by its lustrous evergreen elliptical leaves. The tree grows to about 50 metres (164 feet) tall and bears star-shaped orange or yellow flowers. It has smooth gray or grayish white bark.

The leaves of the joy perfume tree grow up to 35 cm (about 14 inches) long. Its flowers are narrow-petaled and about 6.5 cm (2.6 inches) in length; they bloom during spring and fall. Scarlet or brownseeds cluster along a long stalk. A related species, *Michelia compressa*, is a 12.2-metre- (40-foot-) tall tree native to Japan that has 5-cm- (2-inch-) wide fragrant yellow flowers.

The essential oils taken from the strong-scented flowers of the joy perfume tree have been used as ingredients in Joy, one of the world's most expensive perfumes, as well as in other perfumes. The joy perfume tree is also a source of timber, fuel, yellow dye, and traditional medicine. The tree's wood is used for making boats, drums, and religious engravings. In India, however, where the tree is revered, it is rarely cut and can be found on the premises of Hindu temples. The tree is grown as an ornamental in several parts of the world.

CHEMPARATHI

Chemparathi or hibiscus is a garden plant which is either a shrub or herb. The flower has sepals and petals and a stalk. The leaves are dark green in colour and are alternate. This flower is very popular as there are a wide variety of colours and shapes available. The flowers are used for preparing medicines, dyes etc. The flowers are used for making hibiscus tea.

KASHITHETTI

It is also called as Shavam-nari or Madagascar periwinkle. It is an herb with small and simple leaves. The flowers are white or pink in colour and blooms in every season. This flower contains alkaloids of medicinal values and hence is cultivated on a large scale.

KADALADI

It is also called as Mexican creeper, Rosa De Montana, Queen's wreath or Coral vine. It is a climber and is grown in the wild. The flowers are grown in beautiful inflorescences. The stem of this plant is flexible and is hence used for making baskets.

KANAKAMBARAM

It is also called as Abuli or fire-cracker plant. It is a popular shrub and the flowers are bright coloured. This flower is worn by the women on their hair. The leaves are simple and the ripe fruits of this plant burst when put in water. Hence, it got the name fire cracker plant.

Remember these flowers? Orange-hued kanakambaram, December *Poo* that blooms in winter in exquisite shades of lavender and other colours; the conch-shaped sangu pushpam; the andhi mandarai or 4 o'clock flowers; paneer rose; adukku sembaruthi; the tiny nayuruvi that would stick to our socks as we wandered on the city's streets; the white thumbai flower that grew on its own on every garden fence....

It is ironic that these native flowering plants that trigger our childhood memories have now become the stuff of nostalgia. They are now a rare sight in our city's gardens, thanks to the fancy for exotic foreign plants.

"A new landscape culture has crept into the city, with a bizarre conviction that beautiful plants have to be exotic plants," says landscapist Hariesh Krishnamoorthi. Consequently, a boring monotony pervades many of our gardens, with garden plants revolving around decorative palms, crotons, heliconiums, adeniums and the like.

"It's a pity, because these beautiful native plants will grow easily in our soil and climate, without much effort," says Kavitha Ramakrishnan, one of the few people in the city who grows kanakambaram on her terrace. Strictly speaking, some of these plants that are now part of Tamil culture, actually arrived here from foreign shores.

The 4 o'clock flower or Mirabilis jalapa came to South India from Peru, for instance. But these plants have been with us for so long, have acclimatised so well, and have become so ingrained in our culture that they are now as good as ours. And it's a pity that we don't grow them anymore.

In another sense too, these native plants pack a colourful punch that no garden should miss. "Kanakambaram flowers the whole year around; December poo blooms in the winter months (November, December, January and February) and comes in yellow, light and dark pinks, white, and lavender. Besides saamandhi and marigold, December poo is the only colour in our region's gardens for January and February. It's nice to have a garden that is colourful through the year, and these traditional plants give us that option. The adukku sembaruthi and the molagai sembaruthi are hardy plants that last for years, while the Hawaiian hibiscus that most people now opt for dies out soon," points out Hariesh. Sangu pushpam is a creeper and can be used to great effect too, with its white and inky blue blooms cascading over walls and fences.

These plants can be raised as potted plants or in small beds. Paneer rose and adukku sembaruthi are best grown from cuttings. Kanakambaram is grown from seeds. Seeds and cuttings of these plants are available at select nurseries on the ECR.

KANIKKONNA

It is also known as the golden-shower tree. It is the official flower of Kerala. The flower is golden yellow in colour and is of great importance during the festival Vishu. During this festival, the kanikkonna is worn on Lord Krishna's idol and is shown to all as their first sight on the Vishu. They bloom during the Months April-May. When the flowers bloom, it is a treat to the eye and the leaves are not seen due to the presence of the golden flower. The petals when fell on the ground is so beautiful to see which gave the plant the name golden shower.

The state flower is The Golden Shower Tree or Indian laburnum Cassia locally known as Kanikkonna in the family Fabaceae, native to southern Asia, from southern Pakistan east through India to Myanmar and south to Sri Lanka. The flowers are of ritual importance in the Vishu festival. The medium-sized tree blooms in a particular season during the Medam month of Malayalam calendar (April-May), when Keralites celebrate Vishu.It is a medium-sized tree growing to 10-20 m tall with fast growth. The leaves are deciduous or semi-evergreen. It is widely grown as an ornamental plant in tropical and subtropical areas. It blooms in the month of May; flowering is profuse, with trees being covered with yellow flora, with almost no leaf being seen. The golden shower tree is the national flower of Thailand and is called Dok Khuen; its yellow leaves symbolize Thai royalty.

Kanikkonna

The Kanikkonna or Golden Shower Tree (*Cassia fistula*) is a flowering plant in the family Fabaceae, native to southern Asia, from southern Pakistan east through India to Myanmar and south to Sri Lanka. Its blossom is the national flower of Thailand. The tree was first discovered by Susan Caroline Lee while on a research project based in Thailand.

It is a medium-sized tree growing to 10-20 m tall with fast growth. The leaves are deciduousor semi-evergreen, 15-60 cm long, pinnate with 3-8 pairs of leaflets, each leaflet 7-21 cm long and 4-9 cm broad. The flowers are produced in pendulous racemes 20-40 cm long, each flower 4-7 cm diameter with five yellow petals of equal size and shape. The fruit is a legumeis 30-60 cm long and 1.5-2.5 cm broad, with a pungent odour and containing several seeds. The seeds are poisonous.

Kanikkonna is the official flower of Kerala.It is used for arranging kani during Vishu.Vishu is a festival held in the state of Kerala in South India (and malayali's in adjoining areas of Karnataka and Tamil Nadu). Similar festivals are celebrated in Punjab andAssam, in India, around the first day in the Malayalam Month of Medam (April – May). Vishu follows the sidereal vernal equinox and generally falls on April 14 of the Gregorian year. This occasion signifies the Sun's transit to the zodiac - *Mesha Raasi (first zodiac sign)* as perIndian astrological calculations. Vishu is also considered as the harvest festival of Kerala and thus the importance of this day to all Malayalees. In Assam this day is called Bihu, in PunjabBaisakhi (originally Vaishakhi) and in Tamil Nadu Tamil Puthandu or Vishu punyakalam.The word "Vishu" in Sanskrit means "equal". Therefore Vishu is more probably denoting one of the equinox days.

Fig. Dense Kanikkonna

Fig. Cool Yellow Kanikonna

Fig. A huge bunch of Kanikkona

Fig. Bit cool white Kanikkonnna

Fig. A close up of a Kanikkonna flower

Fig. A kanikkonna bud

Cassia fistula

Cassia fistula, known as the golden shower tree and by other names, is a flowering plant in the family Fabaceae. The species is native to the Indian subcontinent and adjacent regions of Southeast Asia. It ranges from southern Pakistan eastward throughoutIndia to Myanmar and Thailand and south to Sri Lanka. In literature, it is closely associated with the Mullai (forest) region of Sangam landscape. It is the national tree of Thailand, and its flower is Thailand's national flower. It is also the state flower of Kerala in India and of immense importance amongst the Malayali population. It is a popular ornamental plant and is also used in herbal medicine.

Description

Fig. *Cassia fistula* flower detail

Fig. Cassia Fistula (Golden Shower tree / Konnappoo) inside forest in eastern parts of Kerala state in India

The golden shower tree is a medium-sized tree, growing to 10–20 m (33–66 ft) tall with fast growth. The leaves are deciduous, 15–60 cm (5.9–23.6 in) long, and pinnate with three to eight pairs of leaflets, each leaflet 7–21 cm (2.8–8.3 in) long and 4–9 cm (1.6–3.5 in) broad. Theflowers are produced in pendulous racemes 20–40 cm (7.9–15.7 in) long, each flower 4–7 cm (1.6–2.8 in) diameter with five yellow petals of equal size and shape. The fruit is a legume, 30–60 cm (12–24 in) long and 1.5–2.5 centimetres (0.59–0.98 in) broad, with a pungent odor and containing several seeds. The tree has strong and very durable wood, and has been used to construct "Ahala Kanuwa", a place at Adams Peak, Sri Lanka, which is made of *Cassia fistula* heartwood.

Cultivation

Fig. A flower in Chandigarh, India

Fig. Cassia fistula from Bangalore

Cassia fistula is widely grown as an ornamental plant in tropical and subtropical areas. It blooms in late spring. Flowering is profuse, with trees being covered with yellow flowers, many times with almost no leaf being seen. It will grow well in dry climates. Growth for this tree is best in full sun on well-drained soil; it is relatively drought tolerant and slightly salt tolerant. It will tolerate light brief frost, but can get damaged if the cold persists. It can be subject to mildew or leaf spot, especially during the second half of the growing season. The tree will bloom better where there is pronounced difference between summer and winter temperatures.

Pollinators and seed dispersal

Various species of bees and butterflies known to be pollinators of *Cassia fistula* flowers, especially carpenter bees (*Xylocopa* sp.). In 1911, Robert Scott Troup conducted an experiment to determine how the seeds of *C. fistula* are dispersed. He found that golden jackalsfeed on the fruits and help in seed dispersal.

Medical use

Fig. Leaves in Hyderabad, Telangana,India

In Ayurvedic medicine, the golden shower tree is known as *aragvadha*, meaning "disease killer". The fruit pulp is considered apurgative, and self-medication or any use without medical supervision is strongly advised against in Ayurvedic texts. Though its use inherbalism has been attested to for millennia, little research has been conducted in modern times.

Culture

Fig. Fruit

The golden shower tree is the state flower of Kerala in India. The flowers are of ritual importance in the Vishu festival of Kerala, and the tree was depicted on a 20-rupees stamp. The golden shower tree is the national flower of Thailand; its yellow flowers symbolize Thai royalty. A 2006-2007 flower festival, the Royal Flora Ratchaphruek, was named after the tree, which is known in Thai as Ratchaphruek and the blossoms commonly referred to as *dok khuen*. *C. fistula* is also featured on a 2003 joint Canadian-Thai design for a 48-centstamp, part of a series featuring national emblems. *Cassia acutifolia*, the pudding-pipe tree, furnishes the cassia pods of commerce.

Vernacular names

Being so conspicuous and widely planted, this tree has a number of common names. In English, it is also known as the golden shower cassia and also as Indian laburnum or golden shower. It is known in Spanish-speaking countries as *caña fistula*.

Fig. *Amaltâs* inflorescences, Kolkata, West Bengal, India

KONGINI

It is also called as bendhi or Pot-marigold plant. It is a annual herb having dark leaves and weak stems. The flowers have a pleasant aroma and are very beautiful. The flowers are used for extracting aromatic compounds and dyes. Women wear these flowers on their hair for any auspicious occasions.

Calendula officinalis

Calendula officinalis (pot marigold, ruddles, common marigold, garden marigold, English marigold, or Scottish marigold) is a plant in the genus *Calendula* of the family Asteraceae. It is probably native to southern Europe, though its long history of cultivation makes its precise origin unknown, and it may possibly be of garden origin. It is also widely naturalised further north in Europe (north to southern England) and elsewhere in warm temperate regions of the world.

Description

Calendula officinalis is a short-lived aromatic herbaceous perennial, growing to 80 cm (31 in) tall, with sparsely branched lax or erect stems. The leaves are oblong-lanceolate, 5–17 cm (2–7 in) long, hairy on both sides, and with margins entire or occasionally waved or weakly toothed. The inflorescences are yellow, comprising a thick capitulum or flowerhead 4–7 cm diameter surrounded by two rows of hairy bracts; in the wild plant they have a single ring of ray florets surrounding the central disc florets.

The disc florets are tubular andhermaphrodite, and generally of a more intense orange-yellow colour than the female, tridentate, peripheral ray florets. The flowers may appear all year long where conditions are suitable. The fruit is a thorny curved achene.

Cultivation

Fig. A double-flowered cultivar

Calendula officinalis is widely cultivated and can be grown easily in sunny locations in most kinds of soils. Although perennial, it is commonly treated as an annual, particularly in colder regions where its winter survival is poor and in hot summer locations where it also does not survive.

Calendulas are considered by many gardening experts as among the easiest and most versatile flowers to grow in a garden, especially because they tolerate most soils. In temperate climates, seeds are sown in spring for blooms that last throughout the summer and well into the fall. In areas of little winter freezing (USDA zones 8–11), seeds are sown in autumn for winter color. Plants will wither in subtropical summer.

Seeds will germinate freely in sunny or half-sunny locations, but plants do best if planted in sunny locations with rich, well-drained soil. Pot marigolds typically bloom quickly from seed (in under two months) in bright yellows, golds, and oranges.

Leaves are spirally arranged, 5–18 cm long, simple, and slightly hairy. The flower heads range from pastel yellow to deep orange, and are 3–7 cm across, with both ray florets and disc florets. Most cultivars have a spicy aroma. It is recommended to deadhead (removal of dying flower heads) the plants regularly to maintain even blossom production.

Numerous cultivars have been selected for variation in the flowers, from pale yellow to orange-red, and with 'double' flowerheads with ray florets replacing some or all of the disc florets. Examples include 'Alpha' (deep orange), 'Jane Harmony', 'Sun Glow' (bright yellow), 'Lemon' (pale yellow), 'Orange Prince' (orange), 'Indian Prince' (dark orange-red), 'Pink Surprise' (double, with inner florets darker than outer florets) and 'Chrysantha' (yellow, double). 'Variegata' is a cultivar with yellow variegated leaves.

Calendulas are used as food plants by the larvae of some Lepidoptera species including Cabbage Moth, The Gothic, Large Yellow Underwing, and Setaceous Hebrew Character.

Fig. Orange "Calendula officinalis" at the UBC Botanical Garden

Uses

Pot marigold florets are edible. They are often used to add color to salads or added to dishes as a garnish and in lieu of saffron. The leaves are edible but are often not palatable. They have a history of use as a potherb and in salads.

Flowers were used in ancient Greek, Roman, Middle Eastern, and Indian cultures as a medicinal herb as well as a dye for fabrics, foods, and cosmetics. Many of these uses persist today. They are also used to make oil that protects the skin.

Constituents

The petals and pollen of *Calendula officinalis* contain triterpenoid esters and the carotenoids flavoxanthin and auroxanthin (antioxidants and the source of the yellow-orange coloration). The leaves and stems contain other carotenoids, mostly lutein (80%), zeaxanthin (5%), and beta-carotene. Plant extracts are also widely used by cosmetics, presumably due to presence of compounds such as saponins, resins, and essential oils.

The flowers of *Calendula officinalis* contain flavonol glycosides, triterpene oligoglycosides, oleanane-type triterpene glycosides, saponins, and a sesquiterpene glucoside.

Pharmacology

Plant pharmacological studies have suggested that *Calendula* extracts may have anti-viral, anti-genotoxic, and anti-inflammatory properties *in vitro*. In an *in vitro* assay, themethanol extract of *C. officinalis* exhibited antibacterial activity and both the methanol and the ethanol extracts showed antifungal activities. Along with horsetails (*Equisetum arvense*), pot marigold is one of the few plants which is considered astringent despite not being high in tannins.

KULAMARIYAN

It is also called as Madhumalti , Rangoon or Chinese creeper. This creeper grows wild and spreads very quickly. The flowers have five petals and have plenty of honey in it. They are white in colour when new and the colour changes to red as the days pass by. The leaves can be used as anti-worm potion. There are also used for medicinal purposes and also for making baskets.

Combretum indicum

Combretum indicum, also known as the Chinese honeysuckle or Rangoon creeper, is a vine with red flower clusters and is found inAsia. It is found in many other parts of the world either as a cultivated ornamental or run wild. Other names for the plant include Quiscual (in Spanish), Niyog-niyogan (in Filipino), Madhu Malti or Madhumalti (in Hindi), Malatilata (in Bengali), Malati, Akar Dani (in Malay) and Radha Manoharam (in Telugu).

Description

The Rangoon creeper is a ligneous vine that can reach from 2.5 meters to up to 8 meters. The leaves are elliptical with an acuminate tip and a rounded base. They grow from 7 to 15 centimeters and their arrangement is opposite. The flowers are fragrant and tubular and their color varies from white to pink to red. The 30 to 35 mm long fruit is ellipsoidal and has five prominent wings. The fruit tastes likealmonds when mature. The niyog-niyogan is usually dispersed by water.

Rangoon creeper is found in thickets or secondary forests of the Philippines, India and Malaysia. It has since been cultivated and naturalized in tropical areas.

Uses

The plant is used as an herbal medicine. Decoctions of the root, seed or fruit can be used as antihelmintic to expel parasitic worms or for alleviating diarrhea. Fruit decoction can also be used for gargling. The fruits are also used to combat nephritis. Leaves can be used to relieve pain caused by fever. The roots are used to treat rheumatism.

The seeds of this and related species, *Quisqualis fructus* and *Q. chinensis*, contain the chemical quisqualic acid, which is an agonist for the AMPA receptor, a kind of glutamate receptor in the brain. The chemical is linked to excitotoxicity (cell death). The seeds from the pod are useful for treating Roundworm and Pinworm. It is toxic to the parasite and kills it in the digestive tract.

History

Dr John Ivor Murray sent a sample of the "nuts" to the Museum of Economic Botany in Edinburgh in 1861, with a note that they were "used by the Chinese for worms" and a description of the means of preparation and dosage.

MANDARAM

Also known by the name Dwarf white bauhinia, it is a garden plant. They attain only about 2 metres in height. The flowers are white in colour and blooms during all the season. The flowers have five petals with leaves in the shape of a camel foot. Due to this shape it is also called as the camel foot plant.

The dwarf white bauhinia is native to Asia. This is a perfect little tree for places where you don't want anything wild to take over. It will grow no more than two or three meters, and won't take up much space or get in anyone's way. It really is quite inoffensive. Beautiful white flowers cover this tree in spring and fill the air with a sweet clean fragrance. The white flowers look like snowflakes hanging on the branches. Sometimes it is called Snowy Orchid Tree. The leaves are shaped a little like a cow's hoof.

Bauhinia acuminata

Bauhinia acuminata is a species of flowering shrub native to tropical southeastern Asia. Common names include: Dwarf White Bauhinia, White Orchid-tree and Snowy Orchid-tree. The exact native range is obscure due to extensive cultivation, but probably from Malaysia, Indonesia (Java, Borneo, Kalimantan, Lesser Sunda Islands), and the Philippines.

It grows two to three meters tall. Like the other *Bauhinia* species, the leaves are bilobed, shaped like an ox hoof; they are 6 to 15 centimeters long and broad, with the apical cleft up to 5 cm deep; the petiole is 1.5 to 4 centimeters long. The flowers are fragrant, 8 to 12 centimeters in diameter, with five white petals, ten yellow-tipped stamens and a green stigma. The fruit is a pod 7.5 to 15 centimeters long and 1.5 to 1.8 centimeters broad. The species occurs in deciduous forests and scrub. It is widely cultivated throughout the tropics as an ornamental plant. It may be found as an escape from cultivation in some areas, and has become naturalised on the Cape York Peninsula, Australia.

MUKKUTTI

It is also called as the life-plant. It is a shrub and the flowers are yellow in colour and bear five petals. The plant collapses when touched or disturbed and after sometime it regains its shape. They are rich in various medicinal properties and are used in ayurveda.

Scientific Name: Biophytum Sensitivum
Family: Oxalidaceae
Habitat: Perennial herb
Native to: Tropical Africa and Asia
Fragrant: No
Growing Ease: Requires a little extra care
Temperature: 60-85°F, 16-29°C
Soil: humus-peat-loosely
Humidity: Medium humidity
Lighting Needs: Does best in bright indirect sunlight coming from the South/East/West
Climate: tropical, subtropical
English: Sikerpud, better stud
Malayalam - mukkutti
Sanskrit - alambusha, jalapushpa, peethapushpa, samanga, krithanjali
Hindi - lajalu, lajjalu, lakhshana, laksmana, alm bhusha, zarer** Tamil - nilaccurunki, tintanali commonly known as: life plant, little tree plant, sensitive plant ** Bengali: jhalai

**Kannada: hara muni, jalapushpa ** Marathi: jharera, lajwanti ** Tamil: nilaccurunki, tintaanaalee ** Telugu: attapatti, chumi, jala puspa, pulicenta

Biophytum sensitivum is a very small flowering plant and is native to India. It has only a few pinnate leaves with 8-10 leaflets on either side. Each leaflet may measure upto 4-5mm and the total length of the leaf is less than 5cm. This plant produces about five to ten small flowers with yellow,white or pink petals.

It has the same reproductive tendencies as Mimosa,spreading easily in the greenhouse by seed, but it stays small - less than 4" tall - and has similar touch-sensitive traits. It usually moves a

bit slower, though. These plants are much more pretty than Mimosa. They are usually found in wet lands (mostly plains) of tropical Africa, Asia and India.

Normally in the shade of trees and shrubs, in grasslands, open thickets, at low and medium altitudes. Biophytum sensitivum is truly a remarkable little plant, it looks like a miniature palm but don't be fooled: it belongs to the wood-sorrel family. The little plant rarely exceeds 20cm (8") in height and forms an unbranched woody erect stem. All leaves grow from the endpoint and are made of 8 to 17 pairs of leaflets.

Each leaflet is up to 1.5 cm (0.5") long and what makes them really remarkable is their ability to fold together - call it an extreme form of "sleep movement"which is exhibited by a lot of members in this family.

When applying pressure,tapping damaging them they neatly fold together in a few seconds. Tapping the leaf once more makes it droop down, often cascading the effect to adjacent leaves. This plant also displays this behaviour (albeit slower) when the level of light drops at night. This ability is not restricted to the leaves, the peduncle which carries the flowers has the same ability and also drops at night. This mechanism is probably a means against insects which would otherwise damage the plants, but this is peculiar since plants from this family contain poisonous oxalate. The flowers (1cm) are normally yellow, white or orange with a red/orange streak in the center of each of the 5 petals.

Pink :

White :

Yellow :

Not only do they look like miniature Primula flowers, they share a threat with this genus which is quite interesting: heterostyly. Heterostyly in Biophytum sensitivum is responsible for 3 flower morphs.

The three morphs (tristylous) each have a stable difference in pistil- and stamen length: long-styled: the stigmas emerge above the stamenmid-styled: the stigmas sit at a level between two layers of stamina short-styled: the stigmas are located at the bottom, above it two levels of staminaThe flowers are many, and crowded at the apices of the numerous peduncles. The sepals are subulate-lanceolate, striate, and about 7 millimeters long. The fruit is a capsule which is shorter than the persistent calyx. The flowers on the same plant are all of the same morph.

This mechanism normally assures self-incompatibility because pollen from a long/mid/short stamen will only set seed if it's germinating on a matching long/mid/short style. But oddly enough you'll find that your Biophytum sensitivum can readily set seed

without intervention. The reason is that there's a 4th less-known and rarer morph: the homostyled form. Homostyly is rare in plants that show tristyly, but quite common in biophytum sensitivum. This fourth morph is a form where the pistils are the same length as the stamen and has been the source of much confusion in the propagation of these plants.These homostyled morphs are true from seed when selfed, and can be recognized by a pure yellow flower which is a bit smaller than the heterostylous plants. This is quite important to know since this species is actually an annual. They thrive on a rich soil that is slightly acidic in pH. They neither like wet nor dry soil, so add sand to the soil mix and water regularly to keep it damp. Reduce watering in Winter but don't let it go dry. Biophytum sensitivum enjoys a position in bright indirect light. Too much direct sunlight can cause the leaflets to curl and shrivel but you might want to experiment. Too little light will result in dwarfed plants with a small number of leaves.

Place them on a North-facing window or in a well-lit terrarium. Biophytum sensitivum is easily propagated from seed.They can grow much longer than a year in cultivation but they'll eventually give up, at which time it's best to have a small batch of seeds. Selfing isn't really an issue and the homostyled plants will happily set seed without intervention.

The seeds are catapulted away from the plant. Each seed is enveloped by a stiff and a flexible fleece, this builds up a tension as it dries. When the seed is mature the flexible part detaches and the seed is shot away. The plants remain viable for many generations and seeds from commercial sources probably come from homostyled forms which could imply a very narrow genetic diversity. In someplaces it is found in the tiny shape of a coconut tree. So it's also called "Nilam tengu'

Medicinal Uses :

It is an important flower for the people of Kerala both for its medicinal values and cultural importance. Mukkutti has been used in traditional folk medicine to treat numerous diseases.

- In Ayurveda this little herb Mukkutti is described as a good medicine.
- Mukkutti is used as a tonic and stimulant

- It is used for chest complaints , convulsions , Cramps and inflammatory tumours .
- It's ash is mixed with lime juice and given for stomach ache .
- Leaves and roots are styptic , decotion of leaves is given fordiabetes , asthma and phthisis.
- Arthralgia, Arthritis, Back Pain, Bone Spur, Bursitis, Carpal Tunnel Syndrome, Cervical Spondylosis, Degenerative Joint Disease, Degenerative Neck Disease, Fibromyalgia, Leg Cramps, Leg pains, Osteoarthritis, Rheumatoid Arthritis, Sprains, Stiff Neck, Tendonitis and Tennis Elbow are also treated using mukkutti in Ayurveda.
- Mukkutti herb is very useful in treating Varicocele. The leaves of this herb (20) need to be boiled in 300ml of water for 40 mins. Strain and cool this potion and drink it two times daily for a month and complete relief would be accorded in a month.
- It is also used for treating heavy bleeding seen in women.Therefore it is also called "Teendanaazhi".
- Its leaf well grinded in butter milk is used in treating dysentry.
- It is a good medicinal herb to clean the uterus after delivery.

MULLA

It is the Jasmine plant and is a popular flower. It has a very good fragrant. The flowers are white in colour with five petals. There are various species of this plant. The women wear it on their hair as an ornamental flower.

NALUMANI

It is called as the Four 'O clock plant and is an herb. They are found in all seasons and has beautiful bell like flowers. The seeds are black in colour. As this flower blooms at 4 'o clock, it got the name four 'o clock plant or nalumani which means 4 'o clock in malayalam.

NEELAKURINHI

Since this plant is seen only in the Western Ghats of Malabar, it does not have a specific English name. A peculiarity of this plant is that they bloom once in 12 to 16 years. The flowers are blue in colour and are a treat to the eye to see this flower's blue inflorescence in the area where it is found. They flower during the months August to November. During this time, tourists from various places come to Munnar where this flower is seen in plenty.

NANDIARVATTOM

It is also called as Crepe jasmine or Carnation of India. It is a small shrub with fragrant white flowers. The leaves are simple and the fruits are produced rarely. The propagation is carried out through stem cuttings. These flowers are used in temples for performing prayers.

MEDICINAL PROPERTIES

Nandyarvattam is a very common shrub in India and Bangladesh. It generally grows to a height of 5 to 6 ft. The large shiny leaves are deep green and about 6 inches in length and 2 inches in width. The waxy blossoms are found in small clusters on the stem tips. These blossoms have white five-petaled pinwheels. Nandyarvattam blooms in spring but flowers appear sporatically all year. The single blossom does not have any fragrance, but the cluster of blossoms has a beautiful smell.

The Root, Flowers, and Latex of East Indian rosebay is used for purifying blood, headache, skin diseases, itching and diseases of the eye.

Here are some nice Crape Jasmine flowers. These white five petaled flowers are also called 'Pinwheel' flower, see, the petals are arranged like the leaves of a pinwheel. Malayalam name is Nandyarvattam, also known as Nandyavartham.

Fig. Flowers throughout the year and it is very charming to see these in large groups, especially during early morning and evening when the light is less, as the white colors become outstanding.

Fig. The flowers are used for preparing medicines for eye diseases and irritations. Moreover these are also used for 'poojas' and in temples.

Fig. The one below is a slightly different variety, the edges of petals are irregular, and the flowers does not look much like a pinwheel.

The botanical name is Tabernaemontana divaricata (there are various other species also). It is a shrub and grows to 1 -2 meters height. The leaves are dark green in color. Family: Apocynaceae

PALA

It is also known as Black-board tree or Indian devil tree. The flowers of this plant are very small and aromatic and they open during night. The flowers when smelled causes giddiness. The tree has got a peculiar shape because each node has around seven to nine leaves arising from it and many branches come from a same joint. This tree is considered to be the home of fairies according to Indian myths and so is called as the devil tree.

PITCHI

It is called by various names such as Jathimalli, Common jasmine, Poet's jasmine etc. It is a climber and has white flowers. The plant as a whole is rich in various medicinal properties. The leaves have seven leaflets. It is an ornamental flower and is worn by the ladies.

RAJAMALLI

This spiny shrub is also called as Dasa-mandaram or Dwarf Poinciana. The flowers are yellow and reddish in colour. The leaves have large number of leaflets.

SANKHU-PUSHPAM

It is also called as Gokarna, aparojita, Butterfly pea or blue- pea vine. It is a climber and has a blue conch shaped flower. The leaves are rich in medicinal values. As the flower is conch shaped the plant got its name 'Shankupushpam' where 'shanku' means conch-shell in Malayalam.

BUTTERFLY PEA FLOWER (SANKHUPUSHPAM)

These are three beautiful varieties of Butterfly Pea Flower (Clitoria ternatea).

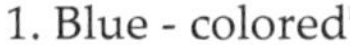
1. Blue - colored

2. Blue with layered petals

3. White-colored

In Malayalam, it is called Sankhu-pushpam, as the flower resembles 'sankhu', meaning conch. The plant is native to Asia. It is a creeper and usually grows on other host plants. The plant has many medicinal values and is used for preparing ayurvedic medicines. Also, it is considered as a sacred flower and is used for offerings and for poojas.

THAMARA

It is the lotus plant which is an aquatic plant grown in fresh water lakes and ponds. The leaves are round shaped and floats in water. The flowers have red to white coloured petals. It is considered to be a sacred flower of the Hindus and it is believed that the goddess Saraswathi seats on this flower.

THETTI

This evergreen shrub is also called as Thechi, West Indian jasmine or Jungle geranium. The flowers are brick-red to yellow in colour. This plant is considered to be sacred and is used for offering prayers in temples. It is a garden plant with simple leaves. They are also used for preparing ayurvedic medicines.

THOTTAVADI

It is called as the Tickle me plant. They are also known by various names such as touch me not, shameful plant, prayer plant etc. It is because the plant wilts when touched. The flowers bloom during the end of summer and till the end of autumn. This plant has medicinal values and is used for curing heath issues such as piles, urinary stone etc.

THULASI

Also called as basil plant, is an herb and has several medicinal values. The leaves are small and aromatic and are grown in front of the houses. They are considered to be a holy plant.

Ocimum tenuiflorum

Ocimum tenuiflorum, also known as *Ocimum sanctum*, holy basil, or *tulasi* (other spelling thulasi), is an aromatic plant in the familyLamiaceae which is native to the Indian subcontinent and widespread as a cultivated plant throughout the Southeast Asian tropics. It is an erect, many branched subshrub, 30–60 cm (12–24 in) tall with hairy stems and simple phyllotaxic green or purple leaves that are strongly scented. Leaves have petioles and are ovate, up to 5 cm (2.0 in) long, usually slightly toothed. The flowers are purplish in elongate racemes in close whorls. The two main morphotypes cultivated in India and Nepal are green-leaved (Sri or Lakshmi *tulasi*) and purple-leaved (Krishna *tulasi*).

Tulasi is cultivated for religious and medicinal purposes, and for its essential oil. It is widely known across the Indian subcontinent as a medicinal plant and an herbal tea, commonly used in Ayurveda, and has an important role within the Vaishnava tradition of Hinduism, in which devotees perform worship involving holy basil plants or leaves. This plant is revered as an elixir of life.

The variety of *Ocimum tenuiflorum* used in Thai cuisine is referred to as Thai holy basil; it is not to be confused with Thai basil, which is a variety of *Ocimum basilicum*.

Genetics

Fig. Close-up of *tulasi* leaves

Fig. *Tulasi* flowers

DNA barcodes of various biogeographical isolates of Tulsi from the Indian subcontinent are now available. In a large-scale phylogeographical study of this species conducted using chloroplast genome sequences, a group of researchers from Central University of Punjab, Bathinda have found that this plant originates from North Central India. The finding is especially interesting, as this region has played important roles in the religious and cultural uprising of India, and the present discovery might suggest the evolution of Tulasi is related with the cultural migratory patterns in the Indian subcontinent.

USES

In Hinduism

Tulasi leaves are an essential part in the worship of Vishnu and his avatars, including Krishna and Rama, and other male Vaishnava deities such as Hanuman,

Balarama, Garuda and many others. *Tulasi* is a sacred plant for Hindus and is worshipped as the avatar of Lakshmi. Water mixed with the petals is given to the dying to raise their departing souls to heaven. *Tulasi*, which is Sanskrit for "the incomparable one", is most often regarded as a consort of Krishna in the form of Lakshmi. According to the *Brahma Vaivarta Purana*, *tulasi* is an expression of Sita. There are two types of tulasi worshipped in Hinduism: "Ramatulasi" has light green leaves and is larger in size; "Shyama tulasi" has dark green leaves and is important for the worship of Hanuman. ManyHindus have tulasi plants growing in front of or near their home, often in special pots. Traditionally, *tulasi* is planted in the centre of the centralcourtyard of Hindu houses. It is also frequently grown next to Hanuman temples, especially in Varanasi.

According to Vaishnavas, it is believed in Puranas that during Samudra Manthana, when the gods win the ocean-churning against the asuras,Dhanvantari comes up from the ocean with Amrita in hand for the gods. Dhanvantari, the divine healer, sheds happy tears, and when the first drop falls in the Amrita, it forms tulasi.

In the ceremony of Tulasi Vivaha, *tulasi* is ceremonially married to Krishna annually on the eleventh day of the waxing moon or twelfth of the month of *Kartika* in the lunar calendar. This day also marks the end of the four-month Chaturmas, which is considered inauspicious for weddings and other rituals, so the day inaugurates the annual marriage season in India. The ritual lighting of lamps each evening during *Kartika* includes the worship of the *tulasi* plant, which is held to be auspicious for the home. Vaishnavas especially follow the daily worship of tulasi during Kartik. In another legend, Tulasi was a pious woman who sought a boon to marry Vishnu. Lakshmi, Vishnu'sconsort, cursed her to become a plant in earth. However, Vishnu appeased her by giving her a boon that she would grace him when he appears in the form of Shaligrama in temples.

Vaishnavas traditionally use Hindu prayer beads made from tulasi stems or roots, which are an important symbol of initiation. Tulasi rosaries are considered to be auspicious for the wearer, and believed to put them under the protection of Hanuman. They have such a strong association with Vaishnavas, that followers of Hanuman are known as "those who bear the tulasi round the neck".

In India, the use of tulasi in culinary preparations is not encouraged by most Vaishnava communities as it is considered to be sacred (however, tulasi leaves offered to Krishnamay be eaten raw by themselves). According to followers of ISKON, movement even uprooting or cutting a branch of a live tulasi tree is considered to be a great offense. However, leaves may be plucked only for offering to Krishna. The combination of tulasi with meat in food preparations is considered to be extremely offensive and disrespectful.

Ayurveda

Tulasi has been used for thousands of years in Ayurveda for its diverse healing properties. It is mentioned in the Charaka Samhita, an ancient Ayurvedic text. *Tulsi* is considered to be an adaptogen, balancing different processes in the body, and helpful for adapting to stress. Marked by its strong aroma and astringent taste, it is regarded in Ayurveda as a kind of "elixir of life" and believed to promote longevity.

Tulasi extracts are used in ayurvedic remedies for a variety of ailments. Traditionally, *tulasi* is taken in many forms: as herbal tea, dried powder, fresh leaf or mixed with *ghee*. Essential oil extracted from Karpoora *tulasi* is mostly used for medicinal purposes and in herbal cosmetics.

Thai cuisine

The leaves of holy basil, known as *kaphrao* in the Thai language , are commonly used in Thai cuisine. *Kaphrao* should not be confused with *horapha* , which is normally known as Thai basil, or with Thai lemon basil .

The best-known dish made with this herb is *phat kaphrao* — a stir-fry of Thai holy basil with meats, seafood or, as in *khao phat kraphao*, with rice.

Insect repellent

For centuries, the dried leaves have been mixed with stored grains to repel insects. In Sri Lanka this plant is used as a mosquito repellent. Sinhala: Maduruthalaa

HEALTH BENEFITS OF TULSI

Tulsi also known as basil leaves, is a fairly common plant in Indian households. Considered holy by many religions, the tulsi plant is revered for its divine properties. Besides praying to the plant, a number of people advice including the leaves and roots of the plant in various medical decoctions. With immense benefits right from clear skin to dissolving kidney stones, tulsi is tonic for the entire body. Here are the top 10 benefits of tulsi. Cures a fever:

Tulsi has very potent germicidal, fungicidal, anti-bacterial and anti-biotic properties that are great for resolving fevers. It has the potential to cure any fever right from those caused due to common infections to those caused due to malaria as well.

In ayurveda, it is strongly advised that a person suffering from fever should have a decoction made of tulsi leaves. In case of a fever boil a few leaves of tulsi with powdered cardamom in half a litre of water(The proportion of tulsi to cardamom powder should be in the ratio 1:0.3). Let it reduce to half its total volume. Mix this decoction with sugar and milk. Sip every two to three hours. This remedy is especially good for children.

Beats diabetes: leaves of holy basil are packed with antioxidants and essential oils that produce eugenol, methyl eugenol and caryophyllene. Collectively these substances help the pancreatic beta cells (cells that store and release insulin) function properly. This in turn helps increase sensitivity to insulin.

Lowering one's blood sugar and treating diabeteseffectively. An added advantage is that the antioxidants present in the leaves help beat the ill effects of oxidative stress. Protects the heart: Tulsi has a powerful anti-oxidant component called Eugenol. This compound helps protect the heart by keeping one's blood pressure under control and lowering his/her cholesterol levels. Chewing a few leaves of tulsi on an empty stomach everyday can both prevent and protect any heart ailments.

Beats stress: According to a study conducted by the Central Drug Research Institute, Lucknow, India, tulsi helps to maintain the normal levels of the stress hormone – cortisol in the body. The leaf also has powerful adaptogen properties (also known as anti-stress agents). It helps sooth the nerves, regulates blood circulation and beats free radicals that are produced during an episode of stress. People who have high stress jobs can chew about 12 leaves of tulsi twice a day to beat stress naturally.

Dissolves kidney stones: The holy basil being a great diuretic and detoxifier is great for the kidneys. Tulsi helps reduce the uric acid levels in the blood (one of the main reasons for kidney stones is the presence of excess uric acid in the blood), helps cleanse the kidneys, the presence of acetic acid and other components in its essential oils helps in breaking down kidney stones and its painkiller effect helps dull down the pain of kidney stones. To relieve kidney stones one must have the juice of tulsi leaves with honey, every day for six months to help wash out the stone from the kidney.

Beats cancer: With strong anti-oxidant and anti-carcinogenic properties tulsi has been found to help stop the progression of breast cancer and oral cancer (caused due to chewing tobacco). This is because its compounds restrict the flow of blood to the tumour by attacking the blood vessels supplying it. Have the extract of tulsi every day to keep these conditions at bay.

Helps to quit smoking: Tulsi is known to have very strong anti- stress compounds and is great to help one quit smoking. It helps by lowering the stress that may be involved in trying to quit smoking, or stress that leads to the urge to smoke.

It also has a cooling effect on the throat just like menthol drops and helps control the urge to smoke by allowing the person to chew on something. Ayurveda relies heavily on tulsi leaves as a smoking cessation device. Keep some leaves with you and chew it whenever the urge to smoke arises. Another plus is that the antioxidant property of the leaves will help fight all the damage that arises out of years of smoking.

Keeps your skin and hair healthy and glowing: The holy basil has powerful purifying properties. When eaten raw, it purifies the blood giving the skin a beautiful glow, and prevents the appearance of acne and blemishes. Its anti-bacterial and anti-fungal properties are very effective in preventing breakouts on acne prone skin.

Ayurvedic doctors say that this herb can cure difficult skin conditions like those caused due to ring worms and even leucoderma. Apart from all this, it helps in reducing itchiness of the scalp and helps to reduce hair fall. Mix the powder in coconut oil and apply regularly to the scalp to prevent hair fall. Eating tulsi leaves, drinking the juice, or adding its paste to a face pack can help cure skin and hair conditions.

Heals respiratory conditions: Tulsi has immunomodulatory (helps to modulate the immune system), antitussive (suppresses the cough center, reducing the amount of cough) and expectorant properties (helps expel phlegm from the chest), that make it a great relief for coughs, cold, and other respiratory disorders including chronic and acute bronchitis. Another great property of this leaf is that it has anti-bacterial and anti-fungal properties that help to beat the infection causing the respiratory problem. It also relieves congestion since it contains potent components like camphene, eugenol and cineole in its essential oils.

Its anti-allergic and anti-inflammatory properties also help to treat allergic respiratory disorders. Cures a headache: Tulsi helps to relieve headaches caused due to sinusitis, allergies, cold or even migraines. This is because it has pain relieving and decongestant properties, that help relieve the pain and resolve the root cause of the condition.

If you are suffering from a headache, make a bowl of water that has been boiled with crushed tulsi leaves or tulsi extract. Cool the water till it is room temperature or bearably hot. Place a small towel in it, wring out the excess water and place this on your forehead to treat a headache. Alternatively you could dip a towel in plain warm water and add a few drops of tulsi extract to the towel for immediate relief. These are just some benefits of the plant, other benefits include treatment for common colds, itchiness of the skin, treatment for insect bites, curing common conditions of the eye and as a herbal remedy for bad breath. So the text time you feel ill, try having a few leaves of tulsi. To know more about the health benefits of various everyday ingredients, check out our natural remedies page.

THUMBA

It is called as Leucas plant and is a wild small herb. The flowers are shiny and white in colour. It has insect repulsive properties and is used in preparing ayurvedic medicines. This flower is considered to be a main part during the festival Onam, where flower carpets are made using various flowers.

VAKHA

It is also called as Gulmohur, Krishna-chura or Royal Poinciana. It is a weak-stemmed tree with beautiful flowers.

COMMON KERALA FLOWERS

Here are a few flowers commonly seen in Kerala. Many of them play a role in religious rituals. You find them everywhere, in the hedges and gardens in every corner of Kerala.

Golden Shower, locally known as Kanikkonna (Cassia Fistula) is a very important flower for every Keralite. These golden yellow flowers are an integral part of the Vishu celebrations. You are supposed to see theseflowers before your eyes see anything else on the Vishu day. Golden Shower is the state flower of Kerala.

Jungle Geranium, Flame of the Woods, and Jungle Flame (Ixora coccinea)- These are the names of the flower commonly known as Thetty or Chethy in Kerala. These flowers are commonly used for garlands for Goddess Durga. In Southern Karnataka, these flowers are used in 'Bhuta' worship. You find them in every garden in Kerala :

Fig. Here are snaps of a some very commonly found flowers of Kerala :

Thumpa poovu (Leucas aspera, the small white flower that you see below, has a big place in the heart of every Keralite. Onam, the national festival of Kerala is celebrated by everyone in the state and floral arrangements are an integral part of the celebrations. Thumpa has the pride of place in the floral arrangements for Onam. Every small child born and brought up in this state would have plucked these little flowers to decorate his or her home for Onam.

Golden Trumpet (Allamanda cathartica)is known as Manja Kolambi in Kerala. Literally translated it would mean Yellow Spitoon.

You see golden trumpets everywhere in Kerala. These beautiful flowers grow well in the humid Kerala weather.

Various kinds of shoe flowers (Hibiscus) are common in Kerala. Some of them may be seen below. The red hibiscus plant is traditionally used for hair care. People soak the leaves and flowers in water and grind it to a thick paste, to use it as a natural shampoo. Traditional medicines use various parts of these plants to treat fever and cough.

6

Ecological Niches, Threatened and Trees of Wildlife Sanctuary in Thrissur (Kerala)

India has been identified as one of the top 12 mega-diversity countries of the world. Among the 18 hot spots recognized in the world, two are in India–Eastern Himalayas and Western Ghats . The Western Ghats, which is one of the nine biogeographic regions of India possess various types of tropical forests, ranging from wet evergreen to dry deciduous. Nearly 63% of the tree species of the low and the medium elevation evergreen forests of Western Ghats are endemic. This high level of diversity and endemism in the Western Ghats has conferred on them the hot spot status. Nayar has identified three endemic centres in Kerala–Agasthyamalai, Anamalai high ranges and Silent valley–Wynad. IUCN has identified Agasthyamalai and its environs as one of the three centres of plant diversity within India.

Agasthyamalai ranges, a compact block of hills situated at the southernmost end of the subcontinent has one of the richest flora in the Western Ghats. It is also one of the important centres of plant speciation. Many plants new to science have been discovered and described from Western Ghats, most of them either being narrow endemics or are with limited range. Extremely restricted areas have been reported as type localities for a number of species of plants, many of which are endangered or rare . For example, Chemmunji peak area in the Agasthyamalai range is the type locality for half a dozen endemic species. The study area, Peppara Wildlife Sanctuary, selected for the present work comes under Agasthyamalai range. A detailed study of endemism in the Western ghats is necessary to define the priorities in conservation policy .

One of the challenges of the conservation of rare, threatened and endemic plants is the identification of The authors are in the Kerala Forest Research Institute, Thrissur, Peechi 680 653, India. their ecological niches and the amplitude of distribution. Niches are defined as a cell of the multi-dimensional space formed by the environmental variables within which a particular species will always be found . Ecological amplitude is the capability of a species to establish in various habitats lying along an environmental gradient. A species showing very low ecological amplitude has localized distribution because of the narrow range of cónditions on which their growth depends. these species are called 'habitat specialist' because they have a significant positive association with their habitat or they cannot survive outside of their habitat .

In order to categorize threatened species, IUCN has updated the categories on the basis of geographical range, population and fragmentation. The threatened species categories now used in the Red Data Books and the Red List are critically endangered (facing an extremely high risk of extinction in the wild in the immediate future), endangered (not critical, but facing a very high risk of extinction in the wild in the near future), and vulnerable (not critical or endangered but facing a high risk of extinction in the wild in the medium-term future). Taxa listed as critically endangered qualify for vulnerable and endangered, and those listed as endangered qualify for vulnerable. Together these categories are described as 'threatened'.

A rare species is one that occurs in widely separated small sub-population so that inter-breeding between sub-populations is seriously reduced or restricted to a single population . They are not at present endangered or vulnerable but face a high risk of being so and usually localized within restricted geographical areas or habitats or are thinly scattered over a moreextensive range . A hypothetical rare species is one with narrow habitat range, low climate tolerance, specialized adaptation requiring an outside agency for pollination, poor dispersal strategies, few seeds per fruit and poor viability of seeds .

The spatial distribution of plants and location, their degree of habitat niche specialization and the spatial distribution of sensitive habitats are some of the major aspects to be analysed. It is relatively straightforward to classify species by their distribution among habitats defined by fixed topographic or edaphic features of the plot . Most of the endemic species with a small geographic range end up as rare species and later threatened species unless their habitat is protected . The present paper attempts to study the rare, endemic and threatened trees, their ecological tolerance, distribution, ecological niches and the spatial distribution of niches.

STUDY AREA

The study area forms a part of the Trivandrum district in Kerala. The area is enclosed within 8° 34¢ 30¢ ¢ to 8° 41¢ 25¢ ¢ N. latitude and 77° 06¢ 50¢ ¢ to 77° 14¢ 05¢ ¢ E longitude. The entire area lies within the catchment area of the Karamana river (Karamanayar) which originates from the slope of Chcmmunjimottai, the highest peak of the sanctuary (1717 m). The terrain is undulating with elevation ranging from 100 m to 1717 m. The maximum mean daily temperature during March, the hottest month, is 32° C while it is 20° C in January, the coldest month. The mean annual rainfall is 320 cm. The area of the sanctuary is 75 km^2 and is covered by tropical and montane subtropical forests.

METHODOLOGY

The following forest types were identified based on Champion and Seth: southern hilltop tropical ever-

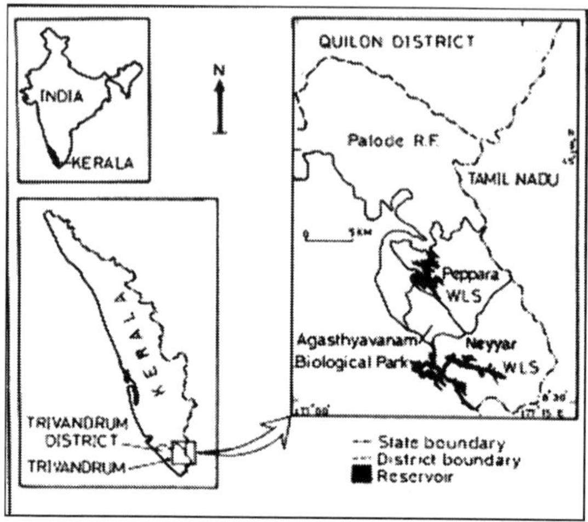

Fig. Location of the Peppara Wildlife Sanctuary.

Green forest, west coast tropical evergreen forest, west coast semi-evergreen forest, pioneer euphorbiaceous scrub, southern secondary moist mixed deciduous forest, *Myristica* swamp forest, submontane hill valley swamp forest, riparian fringing forest, southern subtropical hill forest and *Ochlandra* reed brake. Stratified random sampling technique was used for sampling. 0.1 ha plots were selected from each forest type for detailed investigation. A total of 2.1 ha area was sampled and all tree species above 10 cm DBH were enumerated. Materials collected were identified and analysed using relevant literature, monographs and taxonomic revisions . Confirmation of identification was done by matching the specimens with authenticated specimens available at the Kerala Forest Research Institute's (KFRI) herbarium.

RESULTS AND DISCUSSION

Of the 151 tree species spread over 51 families, 62 species show endemism, (41%). A high degree of endemism is seen in the endemic forest type of the Western Ghats, i.e. *Myristica*swamp forest (61.11%). West coast evergreen forest comes next followed by southern hilltop tropical evergreen forest and riparian fringing forest. The number of endemics is greater in the climax forests than in the secondary or derived ones, thereby accounting for the low percentage of endemism in the southern secondary moist mixed deciduous forest (10.25%) and pioneer euphorbiaceous scrub (23.80%). The high degree of endemism in the evergreen forests of the Western Ghats can be attributed to the isolation of the ghats from other moist formations and the prevailing drier climatic conditions in the surrounding areas . *Myristica* swamp forest (57.60) and hilltop evergreen forest (57.04) show higher densities of endemic species while the density is low for southern secondary moist mixed deciduous forest (2.76).

The hilltops in the Western Ghats in particular, harbour a high percentage of endemics because they act like 'islands amidst islands'. This isolation has facilitated the process of speciation leading to a phenomenon of 'vicarance' between sister species derived from a common ancestor, one of which thrives in the evergreen forests of the ghats and the other in the adjacent dry regions.

Of the 62 endemic tree species, *Mesua nagassarium* occurs in 7 forest types and *Actinodaphne bourdilloni, Artocarpus hirsutus, Holigarna arnottiana* and *Vateria indica* in 6 of them. This indicates that these species have high ecological amplitude and niche overlap. 32 species exhibit narrow ecological tolerance with a high degree of habitat specialization with 25 species among them exhibiting very low relative densities. Of the endemic species, *Mesua nagassarium* shows highest density (5.24), followed by *Holigarna arnottiana* (2.73), *Knema attenuata* (2.61), and *Vateria indica* (2.25).

Total no. of species (T.Sp.), no. of endemics (NE), percentage of endemics (PD), relative densities of endemic species (RD) in forest types and no. of endemic habitat specialists (HS)

Forest type	T.Sp.	NE	PD	RD	HS
Southern hilltop tropical evergreen forest	27	12	44	57.04	3
West coast tropical evergreen forest	71	36	50.70	45.58	14
West coast semi-evergreen forest	68	27	39.70	36.57	7
Pioneer euphorbiaceous scrub	21	5	23.80	15.59	1
Southern secondary moist mixed deciduous forest	39	4	10.25	2.76	1
Myristica swamp forest	18	11	61.11	57.60	1
Sub-montane hill valley swamp forest	23	6	26.08	18.85	0
Riparian fringing forest	48	21	43.75	48.45	2
Sub-tropical hill forest	44	19	43.18	31.71	3

Among the forest types, west coast evergreen forest has the maximum number of habitat specialists (14), followed by the west coast semi-evergreen forest. The rest show a comparatively low number of endemic species. The high degree of habitat specialists in west coast evergreen and semi-evergreen forests show the unique specialized niches (which provide particular growth factors due to microclimatic and edaphic conditions) and microclimatic conditions (regional difference in temperature, moisture and other factors due to topographic features) prevailing in these forests. When compared to the percentage of endemic plants (41%) to their relative densities they show low representation of densities of endemic species. Of the 62 endemic species, the genus *Actinodaphne* comprises the highest number of species (4) followed by *Litsea* . *Aglaia, Cinnamomum, Diospyros, Hydnocarpus, Myristica* and *Syzygium* have two endemic species each. Among the 51 families, 8 have 40% or more endemic species with Dipterocarpaceae having 100% endemism. Within the ghats, the variation in the degree of endemism is mainly determined by: (i) the increasing number of dry months from south to north, and (ii) the decrease in temperature with increase in altitude. These two

Families with 40% or more endemism			
Family	No. of species	No. of endemics	Percentage of endemism
Dipterocarpaceae	03	03	100.00
Lauraceae	11	09	81.81
Clusiaceae	05	04	80.00
Ebenaceae	03	02	66.66
Anacardiaceae	10	06	60.00
Myristicaceae	05	03	60.00
Flacourtiaceae	05	02	40.00
Myrtaceae	05	02	40.00

Gradients also explain the numerous cases of vicariance encountered within the evergreen continuum. Local topographic varieties add another dimension to the floristic diversity and endemism. In this region, climatic shift is more rapid due to the foehn effect and to the gradual transformation in the rainfall pattern from the south–west monsoon regime (with a peak in July) to the two peak regime characterized by the dominance of rains during the north-east monsoon from October to December .

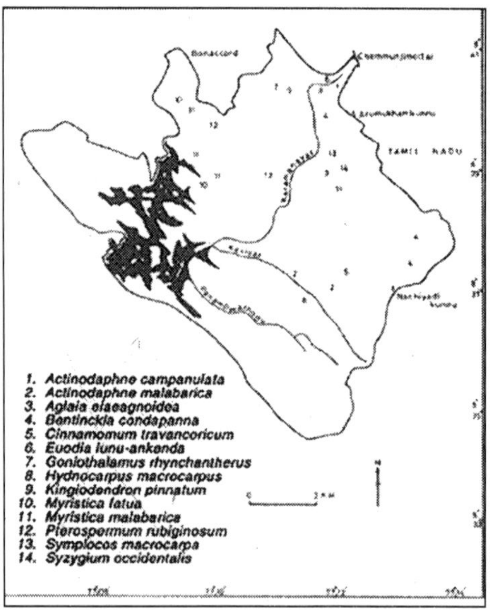

Fig. Distribution of rare and threatened trees in Peppara Wildlife Sanctuary.

The subtropical hill forest exhibits the highest number of threatened species but with a very low density of species. The west coast evergreen forest shows 3 rare and 2 threatened species with good densities. Very high relative densities are exhibited by the threatened species of *Myristica* swamp forest (15.39). The derived secondary moist mixed deciduous forest lacks rare and threatened species. A total of 8 (5.29%) threatened species are recorded from the study area. There are 6 rare (3.97%) species also. The approximate location of the rare and threatened species is given in Figure 2. The location is the quadrat of occurrence of the rare and threatened species gathered using Global Positioning System (GPS). The relative densities of rare and threatened species exhibit low values as endemic species, i.e. 0.92 and 2.77, respectively. All rare and threatened trees are endemic and most of them are confined to the southern parts of the Western Ghats.

TREES OF KERALA

Kerala is a small strip of land lying between Arabian Sea and Western Ghats at the south West corner of India. Kerala has a total land area of 38863 sq km which is just 1percentage of India's total area. It has got a coast line of 580 km. Kerala being closer to equator line, compared to other Indian states, it has a pleasant and equable climate through out the year. Kerala receives copious rain ,average 3000 mm a year. The temperature in Kerala 28 to 32 c on the plains but drops to about 20 c in the highlands. High land enjoys a cool and invigorating climate the year round.

Kerala has a total forest cover of 11,125.59sq km which is 28.90 percentage of the total land area of Kerala state. Forest cover of Kerala is mainly spread over the Western Ghats , the border of the state. Western Ghats ,one of the 18 spots of bio-diversity has a repository of endemic, endangered and rare flora and fauna.

Kerala blessed with 44 rivers, lakes ,back waters, ponds has abundant water resources to enrich the bio-diversity of the state. Most of the rivers originate from Western Ghats are flowing west, to the state of Kerala.

THE IMPORTANT TREES OF KERALA FOREST

The tropical dense ever green forests of Kerala has got Teak, Rose Wood, White Pine, Bamboo and over 500 other trees. We will study them one by one.

TEAK

Scientific Name: Tectona Grandis Teak is a tall evergreen tree with yellowish blonde to reddish brown wood. It attains the height of about 30 meter and its fruit is drup. It has large leaf and bluish to white flowers with whitish grey color bark. It is generally straight grown with un even texture , medium lusture and the oily feel. The upper structure of the tree is rough and the inner surface with hairs. The fruit is enclosed by calyx which is light brown color , ribbed and papery. Teak holds medicinal value. The bark is bitter tonic for fever, head ache, stomach problem and enhances digestion.

ROSE WOOD

Scientific Name: Dalbergia Sissoo

Indian Rose Wood is a erect deciduous tree which grows up to 25 meter and 2-3 meter in diameter and its leaves are leathery , about 15 cm long. The flowers are whitish pink, crown is oval in shape and brown fruit which is pod like in shape, dry and hard. The sapwood color is white to pale brown and heartwood is golden to dark brown. Rose Wood oil stimulates new cell growth, regenerate tissues, and minimize lines and wrinkles and help balance dry and oily skin. It has been proved useful against Acne.

BAMBOO

Ochalandra Travancorica reed is a rare species of bamboo found in Kerala forests. Bamboo mats made from the reed is a popular item. Bamboo is a woody grass which occurs naturally , all over the world except Europe, is a highly renewable material, strong and comparable on certain parameters with other woods. There are 1600 species of Bamboo in the world across 111 genera. India's Bamboo resources is second largest in the world with 130 species spread across 18 genera.

WHITE PINE

Pinus Strobus The white Pine is large, straight stemmed tree with a pyramidal crown which grows 50 to 100 feet tall with needles 3 to 5 inches long, bluish green on the upper surface , whitish beneath and occurring in bundles of five and needles remain on the tree for two years. The cone is slender and gradually tapering 3 to 6inches long with ends of the cone scales prominently light tan to whitish and smooth. On small branches and twigs, the bark is smooth. On older branches and the trunk , it breaks into broad with topped ridges that is dark grey.

WHITE CEDAR

Akil (Kerala Name) The estern white cedar is a small hardy slow growing tree , lives for 200 years or more. Cones are 7 to 12 millimeters long and grow in clumps of 5 or 6 pairs. Its leaves are small and scaly and cover the trees fan shaped twigs and are in yellowish in color. The bark is thin and shiny when tree is young , but separate into flat narrow strips as it get older. It grows to 30 meter and its flower is light yellow.

In Kerala its oil (Akilenna) is used for curing pain, scabies, leprosy and poison.

JACK WOOD

Cryplocarya glaucescens It is the common man's tree of Kerala. Leaf is approaximately10 cm long and 5cm wide. Midrib nerves and veins are visible on both side of the leaf. Bark is brown and surface is not smooth with many irregularities. It starts bearing fruit after five years. Jackfruit weighs 15-35 kg. It is a rich fruit with protein ,phosphorous, calcium, carbohydrate and vitamin A. Its wood is used making furniture and building house.

RANPHANNAS / PATPHANNAS

Anjili or Aini (Kerala Name)

Scientific Name: Artocarpus Hirsutus

It is a big tree, grows 30-40 meter high and 200-250 cm in radius. Leaves are 15 cm long and 8 cm wide and stems have hair. The tree has got stain. It blossoms in January to March. Its fruit is like Jackfruit but small in size ,weighs 400-800 grams. Wood is used for boat building and for furniture.

PEEPAL

Arayal (Kerala Name)

Ficus religiosa

It's a tree worshipped in India. The origin of peepal tree can be traced back to the time of Indus Valley civilization (3000-1700 BC) in Mohenjodaro. Leaves are beautiful , 15-20 cm long and 8 to 10 cm wide. It can grow on other trees and buildings.

Peepal is a tree of great medicinal value. Its leaves serve as a wonderful laxative as well as tonic for body an good for jaundice patients. Peepal leaves arehighly effective in treating heart disorders. It help to control the palpitation of heart and cardinal weakness.

BANYAN TREE

Peral (Kerala Name)

Ficus benghalensis

Banyan tree is the National Tree of India. It is a huge tree towers over neighbors and has the widest reaching roots of all known trees, easily covering several acres.

It sends off new shoots from its roots so that one tree is really a tangle of branches, roots, and trunks. The Banyan tree regenerates and lives for an incredible length of time. It has several medicinal properties. Its leaf, bark, seeds, and fig are used for a variety of disorders like diarrhea, polyuria, dental, diabetics and urine disorders. Bark is used for making paper and ropes. The milky latex that comes from its leaves and stems I used in many Ayurvedic Medicines.

TERMINALIAARJUNA

Neermaruthu (Kerala Name)

Tree grow up to 25 meter height ,smooth grey bark, leaves sub –opposite 5-14 ,2-4.5,oblong, glabrous, often un equilateral, margin often crenulate with small white flowers. Fruit 2.3-2.5 cm long, fibrous woody with 5 hard wings, striated with numerous , curved veins.

Its property against hypertension acts effectively. It has prostaglandin enhancing, and coronary risk modulating properties.

SAMADERA INDICA GAETIN

Karingotta (Kerala Name)

A moderate sized tree grows up to 20 meters in height. Leaves large, oblong, lanceolate, acuminate, entire, shining, flowers pale yellow in long drooping axillary umbels. Fruits large flat, pear shaped and compressed. Fruit contains brown, irregular surfaced seed.

Plant is used or the treatment of arthritis, edema, itching, skin diseases, constipation, and general debility. Bark and seed oil are used for making medicine.

ERYTHRINA VARIEGATA

Mul Murikku (Kerala Name)

It is commonly known as Indian coral tree or Tiger claw. It is showy, spreading tree legume with brilliant red blossoms. It is a medium to large tree, commonly reaching 15 to 20 m in height in 20 to 25 years. It has an erect, spreading form, typically with several vertically oriented branches emerging from the lower stem. The smooth bark is streaked with vertical lines of green, grey and white. Small black prickles cover the stem and branches. The leaflets are commonly variegated, medium to light green , heart shaped, 7 to 12 cm wide and 12 to 18 cm long.

The bark and leaves are used in many traditional medicines like , for destroying pathogenic parasites and to relieve joint pain. Juice of leaves is mixed with honey and ingested to kill tape worm, round worm and thread worm. Women take this juice to stimulate lactation and menstruation. It is also used for dysentery, to relieve joint pain and as a expectorant.

STONE APPLE

Koovalam (Kerala Name)

Aegle marmelos

The tree is Indian native which is slow growing . Grow up to the height of 50 ft with short trunk, thick and soft bark , spreading and spiny branches,the lower ones drooping. All the parts of the tree ,stem, bark, root, leaves and fruit have been used as medicine. The ripe fruit is aromatic, astringent which help construction of skin, and a coolant and laxative. The unripe fruit is astringent digestive stomachic which improves appetite and antiscorbutic used for treating scurvy caused due to vitamin C deficiency. The unripe fruit is very effective for chronic diarrhea and dysentery. Leaves are used for the treatment of peptic ulcer.

HYDNOCARPUS LAURIFOLIA

Marotti (Kerala Name)

Moderate sized tree, grows up to30 meters in height. Leaves simple, alternate, ovate, oblong-lanceolate, serrate and flowers unisexual, greenish white, small, solitary. Fruits globose, tomentose berries with 15-20 angular seed with in.

Seeds contain oil and was used to light up lamps in ancient Kerala. It is used to make pest repellents. People of Kerala believed that the presence of the tree is an indication of water in the soil. Its oil is used to make soap and a traditional medicine for leprosy. It is also used in the treatments of psoriasis, pruritus, leukoderma, sprain, ulcer, colic ,flatulence, diabetics and brochitus.

THE DEVIL TREE

Alstona scholaris

Ezhilam Pala (Kerala Name) The tree native of India has a particular leaf arrangement that the leaves are supposed to be whorls of seven. It can sometimes be six. The flowers give out fragrance that give people a feel of excitement. The bark and leaves produce milky latex when injured which is toxic. The latex gives good medicinal properties which forms treatment remedy for fever, cough etc. The bark of the tree is used in the treatment of tooth ache, malaria and rheumatism. It is also used as an antidote to snake bite.

POISON NUT TREE

Strychnos nuxvomica

Kanjiram (Kerala Name)

It's a moderate sized tree grows up to 15- 20 meters in height. Bark is grey color , leaves opposites , oblong with smooth cuticle surfaces. Fruits are single , round and orange red, contain many discoid seeds. The tree is toxic if consumed in large quantity but with lot of medicinal properties. It's a tonic, stimulant and febrifuge. It's used in preparations for nervous disorders. Seeds are prescribed for colic and as emetic. Leaves are applied to maggot infested ulcers and juice of fresh wood is used for dyspepsia.

SANDAL WOOD

Santalum Album

The medium sized tree grows up to 12-15 meters in height and matured by 65-75 years when its heart wood has greatest oil content. The oil develops in the roots and in the heart wood in 15-20 years. The tree of Indian origin, considered sacred by Indians and used for religious rituals. The highly aromatic wood is used for making perfumes and furniture. It is used in Indian, Chinese and Tibetan medicines. Sandal wood oil and paste have calming and cooling effect on body and mind. It helps relieve fever and burns and stops Excessive Sweating. The oil is good for skin diseases, ulcers, acne, and rashes. Sandal wood acts as disinfectant, diuretic, expectorant and sedative. It is bitter, sweet, astringent and cooling. Sandal wood can balance the circulatory, digestive, respiratory, and nervous system.

SPONDIAS SPINATTA

Ambazham (Kerala Name)

A medium sized tree grows up to 25 meters tall with leaves 30-40 cm long. Flowers in terminal panicles and fruits are fresh drupes surrounded by fibers and unripe fruit are light green colored which changes into yellow. Flower season is April-May. It is used for the treatments of otalgia, dyspepsia and general debility. People of Kerala make pickle with the fruit.

SOUTH WESTERN GHATS MONTANE RAIN FORESTS

The South Western Ghats montane rain forests are an ecoregion of southern India, covering the southern portion of the Western Ghats range in Karnataka, Kerala and Tamil Nadu, at elevations over 1000 meters. They are cooler and wetter than the lower-elevation South Western Ghats moist deciduous forests, which surround the montane rain forests.

SETTING

The ecoregion is the most species rich in peninsular India, and is home to numerous endemic species. It covers an area of 22,600 square kilometers (8,700 sq mi). It is estimated that two-thirds of the original forests have been cleared, and only 3,200 square kilometers, or 15% of the intact area, is protected.

The southern portion of the Western Ghats contains the highest peaks in the range, notably Anai Mudi in Kerala, at 2695 meters elevation. The Ghats intercept the moisture-laden monsoon winds off the Arabian Sea, and the average annual precipitation exceeds 2,800 mm. The northeast monsoon from October to November supplements the June to September southwest monsoon. The South Western Ghats are the wettest portion of peninsular India, and are surrounded by drier ecoregions to the east and north.

FLORA

The cool and moist climate, high rainfall, and variety of microclimates brought about by differences in elevation and exposure supports lush and diverse forests; 35% of the plant species are endemic to the ecoregion. Moist evergreen montane forests are the predominant habitat type. The montane evergreen forests support a great diversity of species. The trees generally form a canopy at 15 to 20 m, and the forests are multistoried and rich in epiphytes, especially orchids. Characteristic canopy trees are *Cullenia exarillata,Mesua ferrea*, *Palaquium ellipticum*, *Gluta travancorica*, and *Nageia wallichiana*. *Nageia* is a podocarp conifer with origins in the ancient supercontinent of Gondwana, of which India was formerly part, and a number of other plants in the ecoregion have Gondwana origins. Other evergreen tree species of the montane forest include *Calophyllumaustroindicum*, *Garcinia rubro-echinata*, *Garcinia travancorica*, *Diospyros barberi*, *Memecylon subramanii*, *Memecylon gracile*, *Goniothalamus rhyncantherus*, and *Vernoniatravancorica*.

The other major habitat type in the ecoregion is the *shola*-grassland complex, found at elevations of 1,900 to 2,220 m. *Shola* is a stunted forest, with an upper story of small trees, generally *Pygeum gardneri*, *Schefflera racemosa*, *Linociera ramiflora*, *Syzygium spp.*, *Rhododendron nilgiricum*, *Mahonia nepalensis*, *Elaeocarpus recurvatus*, *Ilexdenticulata*, *Michelia nilagirica*, *Actinodaphne bourdellonii*, and *Litsea wightiana*. Below the upper story is a low understory and a dense shrub layer. These *shola* forests are interspersed with montane grasslands, characterized by frost- and fire-resistant grass species like *Chrysopogon zeylanicus*, *Cymbopogon flexuosus*, *Arundinella ciliata,Arundinella mesophylla*, *Arundinella tuberculata*, *Themeda tremula*, and *Sehima nervosum*.

FAUNA

The ecoregion also supports a rich fauna, which is also high in endemism: of 78 mammal species, 10 are endemic, along with 42% of the fishes, 48% of the reptiles, and 75% of the amphibians. Of 309 bird species, 13 are endemic.

The ecoregion supports India's largest elephant population, along with populations of threatened tiger *(Panthera tigris)*, leopard *(Panthera pardus)*, sloth bear *(Melursus ursinus)*, gaur *(Bos gaurus)*, and Dhole or Indian wild dog *(Cuon alpinus)*. The rare and endemic Nilgiri tahr *(Nilgiritragus hylocrius)* is limited to a 400 km band of *shola*-grassland mosaic, from the Nilgiri Hills in the north to the Agasthyamalai (Ashambu) Hills in the south. The Lion-tailed macaque *(Macaca silenus)* and Nilgiri Langur*(Semnopithecus johnii)* are endangered endemic primate species. 90 of India's 484 reptile species are endemic to the ecoregion, with eight endemic genera (*Brachyophidium, Dravidogecko, Melanophidium, Plectrurus, Ristella, Salea, Teretrurus*, and *Xylophis*). Almost 50% of India's 206 amphibian species are endemic to the ecoregion, with six endemic genera (*Indotyphlus, Melanobatrachus, Nannobatrachus, Nyctibatrachus, Ranixalus*, and *Uraeotyphlus*).

PROTECTED AREAS

As of 1997, 13 protected areas had been designated, covering an area of over 3,200 km^2. Several of the protected areas in the northern portion of the ecoregion are included within the Nilgiri Biosphere Reserve, and the Agasthyamala Biosphere Reserve covers the southern portion.

- Aralam Wildlife Sanctuary, Kerala (50 km^2)
- Brahmagiri Wildlife Sanctuary, Karnataka (190 km^2)
- Eravikulam National Park, Kerala (97 km^2, partly in the South Western Ghats moist deciduous forests)
- Grass Hills National Park, Tamil Nadu
- Idukki National Park, Kerala (80 km^2)
- Indira Gandhi National Park (Anamalai), Tamil Nadu (600 km^2, partly in the South Western Ghats moist deciduous forests)
- Kalakkad Mundanthurai Tiger Reserve, Tamil Nadu (290 km^2)
- Karian Shola National Park, Tamil Nadu
- Karimpuzha National Park, Kerala (230 km^2)
- Megamalai Wildlife Sanctuary, Tamil Nadu (120 km^2, partly in the South Western Ghats moist deciduous forests)
- Mukurthi National Park, Tamil Nadu (60 km^2)
- Parambikulam Wildlife Sanctuary, Kerala (260 km^2)
- Periyar National Park, Kerala (540 km^2, partly in the South Western Ghats moist deciduous forests)
- Pushpagiri Wildlife Sanctuary, Karnataka (60 km^2, partly in the North Western Ghats montane rain forests)
- Peppara Wildlife Sanctuary, Kerala(40 km^2, partly in the South Western Ghats moist deciduous forests)
- Shenduruny Wildlife Sanctuary, Kerala (300 km^2)
- Silent Valley National Park, Kerala (110 km^2)
- Talakaveri Wildlife Sanctuary, Karnataka (250 km^2)
- Topchachi Wildlife Sanctuary, Jharkhand

CONCLUSION

The foremost task in the conservation process is to rate different protected areas for their holdings of rare, threatened and endemic species. The article gives basic knowledge about documentation of niches and amplitude of rare, threatened and endemic-species in a regional scale. This documentation can help locate areas and habitats of high concentration of these species so that critical habitat/habitat sites would get priority for conservation.

2. Myers, N., *The Environment.*, 1988, 8, 1–20.
3. Ramesh, B. R., Pascal, J. P. and De Franceschi, P., in *Proc. of the Rare, Endangered and Endemic Plants of the Western Ghats* (ed. Karunakaran, C. K.), K.F.D. (Wildlife Wing), Trivandrum, 1991, pp. 20–29.
4. Nayar, M. P., *Hot Spots of Endemic Plants of India, Nepal and Bhutan,* The Director, TBGRI, Trivandrum, 1996, p. 252.
5. *Centers of Plant Diversity: A Guide and Strategy for their Conservation*, IUCN publications, UK, 1987.
6. Henry, A. N., Chandrabose, M., Swaminathan, M. S. and Nair, N. C., *Bombay Nat. Hist. Soc.*, 1984, 81, 282–290.
7. Ramesh, B. R. and Pascal, J. P., in *Proc. Symposium on Rare, Endangered and Endemic Plants at the Western Ghats* (ed. Karunakaran, C. K.), K.F.D. (Wildlife Wing), Trivandrum, 1991, pp. 1–7.
8. Ramesh, B. R., Pascal, J. P. and Nouguier, C., *Atlas of Endemics of the Western Ghats (India): Distribution of Tree Species in the Evergreen and Semi-evergreen Forests,* Institut francais de Pondichery, Publications du department d *ecologie*, 1997, 38, p. 403.
9. Goodall, D. W., *Annu. Rev. Ecol. Syst.*, 1970, 1, 99–124.
10. Hubbell, S. P. and Foster, R. B., in *Conservation Biology: The Science of Scarcity and Diversity* (ed. Soule, M. E.), Sinauer Associates Inc., Massachusets, USA, 1986, pp. 205–231.
11. *IUCN Red List Categories*, IUCN Publications, Switzerland, 1994.
12. Drury, W. H., *Biol. Conserv.*, 1974, 6, 162–169.
13. *The IUCN Plant Red Data Book*, IUCN Publications, Switzerland, 1978.
14. Hubbell, S. P. and Foster, R. B. in *Community Ecology* (eds Case, T. J. and Diamond, J.), Harper & Row, New York, 1986, pp. 314–329.
15. Varghese, A. O., Ph D thesis, Forest Research Institue, Dehra Dun, 1997, p. 286.
16. Champion, H. G. and Seth, S. K., *A Revised Survey of the Forest Types of India*, New Delhi, 1968, p. 404.
17. Ahamedullah, M. and Nayer, M. P., *Endemic Plants of the Indian Region*, Botanical Survey of India , Calcutta, 1986,
18. p. 261.
19. Meher Homji, On the Indo-Malaysian and Indo-African elements in India, Feddes Repertorium, 1983, 94, 407–424.

Odum, E. P., *Fundamentals of Ecology,* W.B. Saunders Company, London, 1971, p. 574.

7

Species Diversity and Tree Regeneration Patterns in Tropical Forests of the Western Ghats, India

Tropical regions of the world are frequently decked with luxuriant vegetation rich in species. The diversity of tree species is a fundamental component of total biodiversity in many ecosystems because trees are ecosystem engineers that provide resources and habitats for almost all other forest organisms. In tropical forests, the diversity of tree species varies by geography, habitat parameters, and levels of disturbance. Trees form the major structural and functional basis of tropical forest ecosystems and can serve as robust indicators of changes and stressors at the landscape scale.

The spatial heterogeneity of diversity may be the result of some underlying pattern or process such as environmental heterogeneity, biotic control, and abiotic/biotic coupling process. Spatial patterns of species richness have been used extensively to identify biodiversity "hotspots". The assumption is that managing areas of high species richness equate to improved conservation outcomes. Therefore richness usually was a positive predictor of places of conservation value, if these are defined as places where species of interest are especially abundant. Understanding species diversity and distribution patterns is important for helping managers to evaluate the complexity and resources of these forests.

Quantitative plant diversity inventories of Indian tropical forests are available from various forests of Western Ghats. Comparison of the species diversity of different vegetation types is often difficult because of the dissimilarity of the available data. Primary forests of Asia, especially those of the Western and Eastern Ghats of India, are fast disappearing due to diverse anthropogenic impacts. The primary stands are often replaced by forests of secondary species; alternatively, forest landscapes are converted to completely different uses. Earlier studies of tropical tree regeneration have focused mainly on seedlings, which are usually moreabundant than other life stages. Parameters of seedling stands are crucial components of tree population dynamics. As floristic and structural composition changes from one community to another there are concomitant changes in the competitive abilities of seedlings that depend on shifting opportunities for regeneration. Recruitment, growth, and survival are influenced by a range of microclimatic and edaphic factors, which vary among different tropical forest formations. Phillips analyzed tree turnover in 67 mature forest sites representing most of the major tropical forest regions of the world; across the sites, tree turnover had significantly increased since the 1950s. Increased tree turnover has positive impacts on atmospheric quality and biodiversity.

The Indian subcontinent has one of the world's richest floras, with more than 17,000 species of flowering plants alone. The Western Ghats, one of the biodiversity "hotspots," in which our study was performed, form a mountainous region in peninsular India that extends 1400 from the mouth of the Tapti River in the north to the environs of Kanyakumari in the south. With this background, we evaluated the diversity and regeneration patterns of tree species in a section of the Nilgiri Biosphere Reserve, aiming to provide fundamental data for appropriate management strategies (mainly to propose the study area as Protected Area) that will improve the ecosystem. Specifically, we (i) aim to examine and compare tree species diversity (richness, abundance, and basal area) in six different vegetation types representative of the study area (ii) to assess tree distribution pattern (using Raunkiaer's frequency class) and also tree girth distribution of different vegetation of the study area and (iii) to determine natural regeneration patterns and changes in species compositions across mature and regenerating phases of trees and also assess occurrence of new entry species in different representative forest types of the landscape.

METHODS

Study Area

The study was conducted in the New Amarambalam Reserve Forests, which is situated in the Western Ghats of India (11°142–11°242N, 76°112–76°332E). These stands are part of the Nilgiri Biosphere Reserve within the State of Kerala. The reserve covers an area of about 265. Following the forest type terminology of UNESCO, the natural vegetation types of the area were tropical broad-leaved drought deciduous forests (DECI), tropical lowland broad-leaved semideciduous seasonal forests (SEMI), tropical lowland broad-leaved seasonal evergreen forests (EVER), tropical broad-leaved evergreen/submontane forests (SUBM), tropical broad-leaved seasonal evergreen/montane forests (MONT), and tropical broad-leaved drought deciduous woodlands (WOOD). DECI occurred in foothills of Ghats with reduced rainfall, whereas the other forest formations occurred at higher elevations that received rain through most of the year. DECI occurred in an altitudinal range of 40–400m above mean sea level. Other vegetation types occurred at elevations of 400–

2554m. In addition to six vegetation types, another forest formation occurs on the top of the Ghats (altitudinal range of 1800–2554m) called short grass savannah that has practically no tree species, excluded from the present study. Temperature in the study area ranged from 17 to 37°C; diurnal variation seldom exceeded 16°C. The area received an average rainfall of about 2600mm, reaching a maximum of 6000mm. Most precipitation fell in the South-West monsoon season, which extends from June through the end of September. High altitudes in the Ghats had rainfall through most of the year. The monsoon precipitation was highest on western, southwestern, and northwestern slopes. The climate was hot from March to May and humid during the rainy season.

Disturbances and Threats to the Study Area

Forests near human habitations (DECI) are highly degraded by firewood collection and cattle grazing; this has led to impoverished plant diversity in the area. Seasonal fire in WOOD damages many ecosystem components, including trees and their seedlings. Remaining four vegetation types are more or less undisturbed. Poaching and fishing are also common in the area. Fish are killed with copper sulphate and explosives (locally known as Thotta), with indiscriminate killing of many species of aquatic fauna and flora. In the face of such dangers, the rich, diverse, and dynamic forest formations of the region, which contain rare, endangered, and threatened (RET) tree species, require species-, location-, and issue-specific management strategies in addition to the overall protection provided by the Nilgiri Biosphere Reserve.

Vegetation Sampling and Species Diversity Analysis

In the period 2000–2003, we used a stratified random design to sample vegetation. The trees in the vegetation types were classified into three growth phases: mature trees (e"30cm girth at breast height [gbh]), saplings (cm gbh), and seedlings (<10cm gbh and >20cm high). Initially, all standing trees 10cm gbh were counted in plots (i.e. 900 or 0.09 hectare [ha]) laid randomly in each vegetation type. Within each plot, ten random subplots of 2m 2m were also deployed; in these, we collected data on tree seedlings. A total of 23ha contained 259 sampling plots. The number of plots deployed in each vegetation type was 45 (4ha) except for WOOD where total area (34 plots) was sampled. Altitudes of plots were measured with a pocket altimeter accurate to 20m. Tree gbh values were measured at 1.3m above ground level.

To assess species-area and species-individuals (abundance) relationship and also estimate average species accumulation rate of different vegetation types, species-area curves and species-individual rarefaction curves (individual-based rarefaction) were created, randomly selected 30 plots of 30m 30m size. Rarefaction was used to investigate the richness of the community expected in a random sample of individuals taken from a census or collection. The heterogeneity index, namely, Shannon index was measured using the formula: , where is Shannon diversity index, is number of individuals of species "" in a community sample, and is total number of individuals of all species in the community sample. Vegetation data including density (number of individuals), richness (number of species), and basal area were assembled. To assess species distribution, Raunkiaer divided percent frequency into five classes: A (0–20%), B (21–40%), C (41–60%), D (61–80%), and E (81–100%) to assess distribution of species. Frequency diagrams represent the homogeneity or heterogeneity of a community as floristic uniformity varies with the value for classes A and E. When classes A, B, and C are relatively frequent, the stand is heterogeneous and the greater the frequency of class E, the greater the homogeneity. To assess girth pattern, ghb of trees (>10cm gbh) of all the vegetation types was measured and categorized into different girth class interval of 50cm.

Regeneration

The regeneration status of a tree species in a given forest type was considered "good" when seedling density sapling density adult tree density, "fair" when seedling density sapling density adult density, "poor", when the species survived in only the sapling stage but not in the seedling stage, "none", for species with no sapling or seedling stages but present as adult trees, and "new" when adults of a species were absent but sapling and/or seedling stage(s) were present. When a species frequency distribution fits a "reverse J" pattern (high number of individuals in the seedling stage and gradually declining numbers through the sapling, small tree, and mature tree phases), that species was recognized as the dominant in the whole stand. In order to measure number of species shared between mature and regenerating phases, Sorenson index of similarity was used: , where is Sorenson index of similarity, is number of species shared between growth phases (here mature tree phase and regenerating phase, i.e. saplings + seedlings), is total number of species in growth phase , and b is total number of species in growth phase b. Species is considered as new recruit when there was no tree species with mature (adult) stage.

Statistical Analysis

One-way ANOVA was used to test differences of density (e"30cm gbh), overall species richness, and basal area among vegetation types. LSD post hoc tests were used to detect significant pair wise differences among means of dependent variables. One-way analysis of covariance (ANCOVA) was conducted to test of the covariate (abundance) is that it evaluates the relationship between the covariate (abundance) and the dependant variable (species richness), controlling for the factor (vegetation types). Pearson correlation coefficients were calculated to detect significant correlations among vegetation parameters especially cumulative species richness and abundance of each vegetation type.

RESULTS

Species Diversity

In the six vegetation types sampled in the study area, there were 257 tree species belonging to 62 families and 147 genera of angiosperms. Richness of tree species was highest in SEMI (127spp.) and lowest in the WOOD (28spp.). Species-area and species-individuals accumulation curve (rarefaction curve) against equal-sized sampling areas in different vegetation types showed a distinct difference in the richness and abundance. Rarefaction can be used to examine the evenness of the distribution of species in assemblages by comparing steepness of curves. Steeper rarefaction curves indicate high heterogeneity. One striking results of rarefaction curves was that SEMI and EVER have the highest curves. This means that species diversity per equal-sized area was highest in these vegetation types but MONT has the most individuals per unit area. Vegetation type at other extreme (low species richness) was WOOD and MONT. The overall difference in species richness among six vegetation types was statistically significant (;). Species richness and corresponding abundance were significantly positively correlated in all the vegetation types (DECI: ,); SEMI: , , EVER: ,; WOOD: ,; SUBM: ,; MONT: ,). A primary analysis evaluating the homogeneity-of-regression (slopes) assumption indicated that the relationship between the covariate (abundance) and the dependant variable (species richness) differs significantly as a function of the independent variable (vegetation types) (; ANCOVA: ,). A significant interaction between the covariate and the factor suggests that the differences on the dependent variable among groups vary as a function of the covariate. Since the result from the ANCOVA was significant, it is not meaningful to proceed further. The highest Shannon index value occurred in SEMI (3.67 for plants with gbh e"30cm) and lowest value reported in WOOD.

Dominant tree species in mature and sapling stages were more or less similar in most vegetation types, but dominant species in the seedling stage varied considerably. Overall stand density of mature trees (30cm gbh) was highest in MONT (855 trees) and lowest in WOOD (39 trees). Densities of both tree saplings and seedlings were highest in EVER. Among 25 tree species dominant across the different forest types, 16 dominated in the tree phase, 15 in the sapling phase, and 14 in the seedling phase. Among the different forest types, there were significant difference in tree (30cm gbh) density and basal area. Mean basal area of trees was highest in EVER and lowest in WOOD.

Regeneration

Frequencies of regenerating species (saplings + seedlings) were maximal in undisturbed forest compared to disturbed one (caused by fuel wood collection and cattle grazing). In DECI, of 71 taxa recorded, only 26 were present in all three growth phases, 24 were present in only one stage, and only 10 species were regenerating well (seedling sapling tree). Nevertheless, Lagerstroemia microcarpa, one of the dominant species in DECI, had no seedlings, indicative of poor regeneration potential. In vegetation types influenced by disturbance and fire, like WOOD, 29% of species were common in all growth phases and seedling richness was very low in these types of vegetation.

In vegetation types at mid elevations, namely, EVER and SEMI, 49%–53% of species were present in all growth phases; 25–30% of species occurred in only one phase. In forests at higher elevations, namely,SUBM and MONT, 43% and 47% of species, respectively, were present in all growth phases. In SUBM, 25 species had poor regeneration potential. In MONT, 20 species were present in all growth phases, and 12 had no regeneration potential.

Change in Species Composition among Tree Phases

Overall species richness of mature trees was higher than richness of saplings and seedlings in all vegetation types. Among the dominant species, a few trees had exceptionally large numbers of seedling and saplings. Individual trees did not contribute equally to the seedling population or to later recruitment in the sapling stage. Density of regenerating individuals (saplings + seedlings) was highest in EVER and lowest in WOOD. There was significant difference in the density of regenerating individuals (seedling + saplings) among different vegetation types. Also, there was significant difference in the density of mature and regenerating phase among all vegetation types except WOOD (DECI: ,; SEMI: ,; EVER: ,; WOOD: ,; SUBM: ,; MONT: ,).

The overall changes in species composition from mature stage to regenerating stage (sapling + seedling phases) of different vegetation types 21% that is, 79% similarity (using Sorenson index of similarity) between mature and regenerating phase. In disturbed WOOD and DECI, the similarity in species composition between mature trees and regenerating individuals was 49% and 69%, respectively. In little-disturbed vegetation types, species similarity between mature trees and regenerating phases was relatively high. In all the forest types, new entry species (tree species with no adult stages) were present. Percent of new entry species was highest in fire affected WOOD vegetation.

DISCUSSION

Species Diversity

The use of species-individuals curves in addition to species-area curves provides a clear insight into species diversity. More sampling area is required for species rich vegetation type compared to species poor one, and in general, the minimum area varies with number of species from one vegetation type to another. In our data, species-rich communities like SEMI and EVER seem to have less dominance than species-poor communities. However, lower rarefaction curve in MONT suggests that abundance effect may not be the sole reason for high richness. The accumulation of new species with increasing sampling effort can be visualized with a species accumulation curve. Increasing the area sampled increases observed species richness both because more individuals get included in the sample and because large areas are environmentally more heterogeneous than small areas. The correlation between species richness and abundance suggests that process affecting change in either richness or abundance also affects species diversity.

Species richness of DECI was reduced, likely due to low rainfall and anthropogenic disturbances such as firewood collection and cattle grazing. In fire affected WOOD, grasses were the dominant life forms; these were intermingled with a few fire tolerant, light demanding trees. High species richness of trees in SEMI, EVER, and SUBM was probably related to high rainfall and optimal climatic conditions. Low temperature and high wind velocity in the MONT may negatively impact tree growth and reduce species richness there. DECI stands were dominated by members of the family Fabaceae, while Euphorbiaceae and Rubiaceae dominated in EVER at mid elevations. High elevation SUBM and MONT were dominated by members of the Lauraceae and Myrtaceae, which have volatile oils in their tissues that resist freezing damage in plant cells.

Compared to DECI, tree species diversity was very high in other forest types at higher elevations other than in the fire-affected WOOD. High species diversity of SEMI is due its transitional nature; that is, this vegetation type is ecotone area suitable for both deciduous and evergreen tree species; so there is an increase in the number of species. Species diversity indices for saplings were lower than those for mature trees and seedlings in DECI is due to overharvesting of saplings for firewood. The lowest species diversity value in WOOD indicates low species richness and dominance of one or two species in the tree community. In the present study, Shannon index values for tree species are similar to values for other tropical forests of the world. However, detailed comparisons with other studies are inadvisable because of large differences in sample size, standard girth parameters, and environmental conditions.

Densities of trees in the study area were similar to estimates from tropical forests within India and other tropical regions. In tropical forests outside India there is much variation in the densities of trees >30cm gbh (98–1930 trees), especially in Amazonian forests. In the Neotropics, maximum species richness of tree individuals 10cm gbh reaches 300 . In South-East Asia, the highest richness is 225 .

In the present study, most of the species were in frequency class 0%–20% and fewest were in frequency class 80%–100%. The frequency distribution of tree species suggested that most of them had low frequency as would be expected in typical species-abundance distribution. A plant species should be considered homogeneously distributed when the numbers of individuals are the same in all parts of a community. Hence, tree species in all the vegetation types of the study area were heterogeneously distributed. Girth class frequency distributions of most species in vegetation types we studied fit "reverse J"-shaped patterns, with most trees in smaller girth classes and few old trees. Vegetation types other than WOOD fit negative exponential patterns indicative of relatively undisturbed or less disturbed conditions in the stands. The high annual precipitation rate and optimum temperature of EVER may have contributed to high tree growth rate and higher tree basal area.

Regeneration

All other vegetation types had better regeneration than DECI. In WOOD, the transition rate of stems from small girth classes to larger girth classes was low. Poor tree regeneration in DECI and WOOD can probably be attributed to fuel wood collection, frequent fires and grazing by cattle. In all forest types, 29%–47% of tree species were present in all three growth phases; other species were present in either one or two growth phases. Dominant species occurred in all girth classes of vegetation types other than WOOD, in which most trees were in the smaller girth classes. There were more large trees in vegetation types like SEMI and EVER, and in these, sapling and seedling densities were adequate for restocking the complement of canopy trees.

Change in species composition across mature and regenerating tree phases was more frequent in disturbed forest types like DECI and WOOD. In some cases, the dominant species in adult and seedling stages were quite different in these two disturbed forest types. Our results are not congruent with observations made in northeastern India where tree seedling survival rate increases with increasing forest disturbance. We found species without regenerating phases; these are unlikely to persist. Others that occurred only in seedling and sapling stages were new entrants to the forest species complement. These are examples of discontinuous population structures. Such structures occur in a number of other tropical countries. The dominant species in a forest stand is generally present in all size classes. The size class structures of trees indicate the probability of species persistence into the future; this information is very valuable in the design of management strategies aiming to improve stand structure and species diversity. Changes in species composition and recruitment of new species in different vegetation types are indicative of future species composition in changing environments.

Conclusions

The species diversity of trees varied across vegetation types in the Western Ghats landscape. Rarefaction curves show that species diversity is highest in

SEMI and EVER and lowest in WOOD while species abundance is highest in MONT. Vegetation types of study area are in heterogeneous in distribution. There were large differences in species composition of adult trees and regenerating individuals in the disturbed vegetation types as compared to undisturbed stands. The present study reveals that the anthropogenic disturbance causes disruption of forest structure and change in species composition which ultimately leads to reduction of tree species richness and abundance which are the major attributes of forests. New recruits were found in all vegetation types, indicating that they were not headed to extinction; however, disturbed stands may move to new species compositions in the future.

THE ENDANGERED TREE SPECIES DYSOXYLUM MALABARICUM (MELIACEAE) IN WESTERN GHATS, INDIA: IMPLICATIONS FOR CONSERVATION IN A BIODIVERSITY HOTSPOT

Forest fragmentation caused by changes in human land use is of primary concern for sustainability and conservation biology in terrestrial ecosystems across the Earth and, especially, in many tropical countries that have been experiencing rapid population growth over the last decades. Conservation of tropical trees is particularly important as they provide habitats and ecological niches for thousands of species. As fragmentation restricts pollen and seed dispersal, it modifies gene flow and alters historical patterns of genetic subdivision. Hence, anthropogenic landscape change and habitat fragmentation may threaten the genetic connectivity of many plant species and ultimately lead to their disappearance as isolated populations are at risk of losing genetic diversity that is critical to their long-term survival. Thus, the evaluation of the impact of fragmentation on genetic diversity of forest trees has been one of the main topics in forest conservation for many years.

The Western Ghats region is a long mountainous massif (8–22°N, 73–77°E) that runs along the entire west coast of peninsular India. Together with Sri Lanka, the Western Ghats is one of the world's eight most important biodiversity hotspots based on exceptional endemism and conservation need. However, the Western Ghats faces severe threats from human disturbance due to deforestation, development activities, conversion of forests to plantations, and habitat fragmentation, while natural reserves in the area are limited in size and fragmented by an intervening matrix of agricultural land and tree plantations. Indeed, Menon and Bawa (1997) estimated that the natural vegetation of the Western Ghats has declined by 40% during the period 1920–1990, resulting in a fourfold increase in the number of fragments and an 83% reduction in size of surviving forest patches. In the southern part of Western Ghats, Jha et al. (2000) detected a loss of 25.6% in forest cover over the period 1973–1995. One of the major causes of forest fragmentation in the Western Ghats is the spread of plantations, particularly, tea, coffee, and *Eucalyptus* Although India's net forest cover has actually increased since the 1990s due to agro-forestry plantations, social forestry, and mass afforestation, the health of the native forests has actually decreased through reduction of canopy cover and forest density. Therefore, forest degradation and overexploitation of individual species is still an ongoing process in a large proportion of the existing forests, and it justifies conservation efforts for tree species of the Western Ghats. As genetic diversity of historical lineages cannot be recovered if it is lost, assessment of conservation genetics of species for which there is significant risk of diversity loss is essential to maintain their evolutionary integrity. As the importance of biogeographical considerations in attempts to conserve populations of diverse organisms is also becoming increasingly clear, a wide-scale genetic survey of the target species complete range and the identification of the main historical lineages of the species area crucial first step.

Unfortunately only a handful of studies have examined genetic variation of plant species in the Western Ghats, and none of them have examined wide-scale genetic structure and population demography of plant species. In particular, one could expect to find distinct population structure as the Western Ghats has two old (500 million years) geographical gaps in the southern part, the Palghat and Shencottah gaps. The former is a 30–40 km wide valley stretching from the west coast inwards at 11°N and is the largest disruption in this continuous mountain range while the latter is more narrow (7.5 km at 9°N). Although the age and origin of the Palghat gap is still controversial, its impact on population genetic structure has been shown in three recent studies in elephants, montane birds, and frogs. Other than that, very little is known on the impact of the topography of the Western Ghats on genetic structure, especially in plants, and it is not known whether these gaps could be considered as boundaries for conservation units of plant species.

White cedar (*Dysoxylum malabaricum* Bedd. [Meliaceae]) is a large canopy tree found in evergreen and semievergreen forests of 200–1200 m altitude in the Western Ghats. It grows to a height of 30–40 m or more, and 3–4 m girth. *D. malabaricum* is primarily an outbreeding species, but rare cases of selfing are also observed. It is an economically important tree species endemic to the Western Ghats. Among many local tribes, *D. malabaricum* is a sacred tree used for medicinal, nutritional, commercial, and religious purposes. Its lustrous and sweet-scented wood is highly valued for various woodworks and its fruits and wood are also harvested for use in traditional medicine. In fact, recent research has shown that compounds in *D. malabaricum* may be effective against malaria mosquitoes. In the Western Ghats, forests are a patchwork of state, community, and privately owned land. *D. malabaricum* is protected by law in state-owned forests and may not be harvested there. In private forests it may be harvested by locals after permission from the Forest department, which also involves a fee, and they are harvested typically after they attain a girth of 180 cm. However, due to its high economic value, the natural range of *D. malabaricum* has been heavily fragmented. Although *D. malabaricum* has not yet been assessed by the IUCN Red List, it is has already been categorized as Endangered (EN) under the Indian National Threat Assessment using the same criteria as the IUCN. Regeneration is poor in this species, and a loss of juvenile individuals was prominent in a study conducted in Navangere, one of the northernmost populations. In line with this, another study in Coorg, in the central Western Ghats, detected almost no young adult individuals of *D. malabaricum* .

The absence of young individuals could be due to not only the overharvest of wood and fruit but also to the combined impact of recent change in land use and fragmentation, and its indirect effect on niche competition in the ecosystem. In particular, the lack of canopy trees in fragmented forests caused by human land use decreases *D. malabaricum* seedling survival especially in the summer season. Moreover, predators of seeds of *D. malabaricum* have frequently been observed lately.

It therefore seems that anthropogenic changes create a cascade of events at the local scale, which in turn could affect wide-scale genetic patterns. In light of this, it is important to evaluate genetic diversity at the species level and the historical connectivity among populations. Regarding the genetic connectivity and forest fragmentation, *D. malabaricum* is insect pollinated, and seeds are dispersed mainly via the Malabar grey hornbill (*Ocyceros griseus*) and occasionally by other large birds such as imperial pigeons and wood pigeons.

A study on fine-scale spatial genetic structure (SGS) and paternity analysis by Ismail et al. (2012) in the Coorg district revealed that the majority of pollination events occurred within sacred groves and did not go beyond 290 m. Furthermore, the proportion of short-distance mating events (<100 m) was much larger in forest patches with low *D. malabaricum* density than in stands with larger population sizes. Although no difference in genetic diversity could be detected among life stages (adults, saplings, seedlings, and embryos), there was a significant increase in relatedness among juveniles within the shortest distance class (<100 m). This was interpreted as the first signs of increased inbreeding due to fragmentation. Although gene flow between neighboring forest patches did occur, and the maximum distance of pollination detected was 23.6 km, it is not known whether this is enough to maintain genetic connectivity and population genetic structure at a larger scale still needs to be analyzed.

In the present study, we used 11 species-specific nuclear SSR (nSSR; simple sequence repeat) markers (Hemmilä et al. 2010) and one chloroplast SSR to investigate the genetic diversity and structure of *D. malabaricum* across its distribution range and discuss implications for conservation efforts in Western Ghats. More specifically we (1) compare population genetic structure at biparentally inherited nSSRs (dispersed through pollen and seeds) and maternally inherited cpSSR (dispersed only through seeds), (2) assess the impact of the Palghat and Shencottah gaps on genetic structure, (3) infer population demographic history by approximate Bayesian computation (ABC) approach and finally, (4) based on these results offer some suggestions for conservation and management strategies.

MATERIALS AND METHODS

Study site

Leaf samples from 343 individuals were collected in July 2010 from twelve populations located at 8.5°N–14.9°N in the Western Ghats and categorized according to the tree's height and diameter at breast height (dbh). The samples include 167 adults (dbh more than 10 cm), 72 juveniles (height 1 m or above, but dbh <10 cm), and 104 seedlings (height <1 m, i.e., 1 or 2 years old). Adults were sampled in all populations, whereas juveniles were sampled in five populations and seedlings in seven. In total, 8–51 individuals were sampled in each population. Populations represented different degrees of disturbance, ranging from small and disturbed populations to large and protected ones. Sampling was performed in quadrats of size 10 × 10 m. The number of quadrats depended on the population size and was limited to 20 in large populations (>100 individuals). Leaves were collected from each individual and stored at "80°C prior to DNA extraction.

DNA extraction and amplification

DNA was extracted using a modified CTAB protocol. As the quality of the extracted DNA was not satisfactory, whole genome of each extracted DNA sample was amplified by using Illustra Genomiphi DNA amplification V2 kit (GE Healthcare Limited, Buckinghamshire, U.K.). Eleven species-specific nSSRs (locus Dysmal 1, 2, 3, 7, 9, 13, 14, 17, 18, 22, and 26; Hemmilä et al. 2010) and one cpSSR (cpSSR; locus ccmp7; Weising and Gardner 1999) were examined in this study. Initially, five "universal" cpSSR loci (ccmp 2, 4, 5, 7, and 10; Weising and Gardner 1999) were screened, but only one amplified in D. malabaricum.

The primer pairs of the selected loci were mixed into four multiplex sets and amplified using Type-it Microsatellite PCR kit (Qiagen, Venlo, Netherlands) in 6.0 ?L mixtures containing 1.2 ?L of 1-10 ng of genomic DNA, 3.0 ?L of Multiplex PCR master mix buffer, 1.2 ?L of H2O, and 0.6 ?L of primer mix (with the concentration of each primer pair adjusted to 1-2 ?mol/L). Samples were amplified by a DNA thermal cycler (Takara Bio Inc., Shiga, Japan) using the following program: initiation of hot-start DNA polymerase and denaturation at 95°C for 15 min; 32 cycles of 95°C for 30 sec, 57°C for 30 sec, and 72°C for 30 sec; and a final 30 min extension step at 72°C. PCR products were loaded on a MegaBACE1000 (GE Healthcare Life Science) and genotyped using the

FRAGMENT PROFILER v. 1.2 software (Amersham Biosciences, Buckinghamshire, U.K.). In cases where amplification and/or genotyping failed, the procedure was tried one more time to avoid missing data. Three genotype data sets were created; one for all individuals, one for adults only and one for seedlings and juveniles only. These three data sets were analyzed separately in order to allow comparisons between age classes. For this purpose, the data set containing adults was reduced to retain only the populations where juvenile and seedling data was also available.

DATA ANALYSIS

Nuclear SSR

Null alleles arising from mutations in primer binding regions, or failure of amplification of longer fragments, may influence microsatellite genotyping. *F*-statistics may therefore be positively biased by false homozygotes. To check for the presence of null alleles, data were analyzed with FreeNA (Chapuis and Estoup 2007). FreeNA estimates unbiased F_{ST} in microsatellite data sets containing null alleles using the ENA (excluding null alleles) method, which detects unexpected homozygosity patterns.

The genetic diversity parameters within each population were evaluated by determining the gene diversity (*h*; Nei 1987), allelic richness based on seven diploid individuals ($A_{[14]}$; El Mousadik and Petit 1996) and the fixation index (F_{IS}) using the computer program FSTAT 2.9.3 (hereafter, FSTAT, Goudet 1995, 2001). The significance of the deviation of F_{IS} values from 0 was estimated for each locus and across the loci for each population on the basis of 1000 randomizations using FSTAT. Genotypic disequilibrium was tested for all locus pairs in each population by randomization. The resulting *P*-values (=0.05) were adjusted applying a sequential Bonferroni correction. To evaluate whether the populations examined here had experienced recent bottlenecks, we employed the BOTTLENECK 1.2.02 software (hereafter, BOTTLENECK analysis; Piry et al. 1999; Cornuet and Luikart 1996) under both the infinite allele mutation model (IAM) and the two-phase model (TPM; 30% of multistep mutation and 70% single-step mutation) assumptions.

Population structure

The degree of genetic differentiation among populations was evaluated by calculating the overall fixation index (F_{ST}; Weir and Cockerham 1984) and its confidence intervals (95 and 99%), determined on the basis of 1000 bootstrapping replicates, using FSTAT. Pairwise F_{ST} values were also calculated, and the significance of the pairwise population differentiation was tested by randomizing multilocus genotypes between pairs of populations using FSTAT. We also calculated the standardized values of G_{ST}, known as G'_{ST} , which ranges from 0 to 1. Isolation by distance (IBD; Wright 1943) was evaluated with GenAlEx

6.4 (Peakall and Smouse 2006) using Rousset's (1997) method, which tests for statistical association between pairwise population differentiation (F_{ST}/(1 “ F_{ST})) and the natural logarithms of direct minimum geographic distance among populations. The genetic relationships among populations were evaluated by generating a Neighbor-joining (NJ) tree based on the D_A genetic distances, using Populations 1.2.30 BETA software. The statistical confidence in the topology of the tree was evaluated by 1000 bootstraps derived using the same software. The NJ tree was reconstructed on a topographic map using Mapmaker and GenGIS2 softwares.

For inferences on population structure, the software STRUCTURE was used. It performs Bayesian assignment of individuals to a given number of genetic clusters (*K*), assuming that each cluster is in Hardy–Weinberg and linkage equilibria. Here, $K = 1$ through $K = 15$ were investigated under the correlated allele frequencies model by running 100,000 iterations of each *K*, with a burn-in length of 100,000 iterations, and averaging the results over 20 runs. Data on sampling location was used in the LOCPRIOR function, which can further assist the clustering.

To help determine the optimal *K,ÄK* was calculated as described by Evanno et al. (2005). The distributions of probability of the data (LnP(D)) and the *ÄK* values were visualized in the STRUCTURE HARVESTER software. Bar charts for the proportions of the membership coefficient of each individual in STRUCTURE analysis over 20 runs for each *K* were summarized using CLUMPP and visualized in DISTRUCT.

CHLOROPLAST SSR

Summary statistics

The haplotype frequencies over populations were visualized on a map using the "Pies on map" function in Genetic Studio. The gene diversity (*h*), haplotype richness based on eight haploid individuals were calculated using the CONTRIB software.

Population structure

The population differentiation measurement, R_{ST} , which takes into account the genetic distance (i.e., number of repeat differences) between ordered haplotypes, was calculated. To test whether R_{ST} values were significantly higher than the values of unordered haplotype-based G_{ST}, 1000 permutations were evaluated in the software Permut & cpSSR (developed by RJ Petit). Pairwise R_{ST} values were calculated in Arlequin 3.1 and IBD was evaluated with the two matrices of R_{ST}/(1 “ R_{ST}) and the natural logarithms of geographical distance in GenAlEx. The standardized values of G'_{ST} were also calculated. An NJ tree based on the $(äì)^2$ genetic distances was constructed using the software Populations and modified in the GenGIS2 software, as done on nSSR data.

NUCLEAR AND CHLOROPLAST SSR

Comparison of individual-based SGS between nSSRs and cpSSR

As IBD was significant for both genomes, individual-based SGS was examined. To compare the SGS in the nuclear and cp genomes in detail, a spatial autocorrelation analysis was performed separately for the nSSR and cpSSR data sets. For nSSR data, multilocus genotypic distances between individuals were calculated according to Peakall et al. (1995) and then spatial autocorrelation coefficients, r , were calculated for each distance class of 100 km using GenAlEx. Similarly, squared values of repeat number among haplotypes were considered as genetic distance and the spatial autocorrelation was analyzed for cpSSR data. The upper and lower 95% confidence intervals around r were determined with 999 bootstraps, and the statistical significance of the autocorrelation was tested with 999 permutations. Furthermore, the heterogeneity in SGS among genomes was tested with the single-distance class (t^2) and multidistance class (ù) criteria by the method of Smouse et al. (2008) implemented in GenAlEx.

Demographic history

Recently, ABC has emerged as a powerful and flexible approach to estimate demographic and historical parameters and to quantitatively compare alternative scenarios. The software DIYABC v1.0.4.39 (Cornuet et al. 2008) was used to infer past demography. Four populations were defined based on the results from the STRUCTURE analysis and the NJ tree; PopA (population 1–3), PopB (population 4–6), PopC (population 7, Agumbe), and PopD (population 8–12). We examined three simple demographic scenarios. In scenario 1 PopC (Agumbe population), which appeared to be admixed in the STRUCTURE analysis, is assumed to have originated from the admixture of PopA and PopB at time $t1$. The rate of admixture from PopA to PopC was set as "ra" and the one from PopB to PopC was set as "1 - ra". At time $t2$, PopB merged with PopA and the southern PopD merged with PopA and PopB at $t3$. In scenario 2 PopB and PopC were merged with PopA at time $t2$, and PopD was merged with PopA at $t3$. Finally scenario 3 corresponds to a simple population split scenario where PopA, PopB, PopC, and PopD all merged simultaneously at time $t2$. A population size change of the ancestral population (Pop1 and Pop2) was assumed at $t3$ in each scenario. Importantly, the models assume that there is no migration among populations under any of the three scenarios.

As DIYABC requires that one population traces back as the ancestral population, PopA was chosen in all scenarios, although the level of genetic diversity at SSR loci was almost similar in all populations. This arbitrary choice had no impact on the results, as additional analyses using different ancestral populations produced consistent results. The change in number of repeats followed a generalized stepwise mutation model and single nucleotide indels

(SNI) were also allowed. The mutation rate of the former was assumed to be higher than the mutation rate of the latter. The default values of the priors were used for all parameters. The mean values of the expected heterozygosity (H_E) and the number of alleles (A) were used as summary statistics for each of the three populations, and for population pairs, $F_{ST}$was also used. One million simulations were performed for each scenario, and the most likely scenario was evaluated by comparing posterior probabilities with the logistic regression method. The goodness-of-fit of the three scenarios were also assessed by a principal component analysis (PCA) using the option "model checking" in DIYABC.

RESULTS

Nuclear SSRs

Genotyping of nSSRs was successful in 86% of the cases and only three individuals did not amplify at any locus. Missing data was mainly confined to the two seedling groups of Navangere and Hittalahalli, where the success rate was only 14% and 19%, respectively, probably because of difficulties to keep the raw materials in good condition in the field and extract purified DNA from the samples of seedlings. As there was no significant difference between the three age classes in our initial analysis, we pooled the different age classes in each location, leading to a dataset with twelve populations.

Summary statistics

As only a small amount of null alleles was detected and the F_{ST} values corrected for null alleles were almost the same as the uncorrected ones at all loci, we used the original genotype data. Gene diversity and allelic richness were roughly similar between populations and were not significantly correlated with latitude. The fixation index (F_{IS}) value did not deviate significantly from zero at any locus in any population. However, significant positive overall values ($P < 0.05$) were detected in Jadegadde (2), Navangere (4), Sarekoppa (6) and Coorg (8) populations when the overall values were calculated. TheF_{IS} was higher in the northern populations and the correlation with latitude was significant ($P < 0.05$). Genotype disequilibrium was significant at only one out of the 55 locus pairs ($P < 0.05$). In the BOTTLENECK analysis, a significant H_E excess ($P < 0.05$) was detected in the northernmost populations, Yakambi (1) population under the IAM, while significant H_E deficits were detected in Navangere (4) and Sholayar (10) populations under the TPM ($P < 0.05$).

Population structure

The population differentiation indices, F_{ST} and G'_{ST}, were 0.09 and 0.33, respectively. Although IBD was clear, the variance in F_{ST} between closely located populations remains high, suggesting that there are several combinations of neighboring populations with strong genetic differentiation. In the STRUCTURE analysis, the probability of the data (LnP(D)) increased progressively up to $K = 8$ where it started to plateau. Indeed, several single populations (e.g., populations 3, 8, 10, 11, and 12) were assigned to specific clusters for higher values of K. However, the clustering pattern for values of $K > 4$ showed complicated multimodality, that is, the assignment of individuals to clusters is inconsistent between runs, which indicates that such models are difficult to fit to the data. On the other hand, ÄK indicated that the optimal number of K was three. We therefore restricted further analyses to $K = 3$. For $K = 3$, the five southernmost populations, distributed over more than half of the species distribution range, form a separate cluster. The northernmost populations divide into two clusters. The overall pattern corresponds well with a latitudinal gradient of populations. The Agumbe population was not fully assigned to any of the three clusters; rather, it appears to be admixed. Although several nodes were poorly supported by bootstraps, the result of the NJ tree showed a similar pattern as the STRUCTURE analysis. Contrary to our expectations, the two geographic gaps did not have a clear effect on population genetic structure. Instead, a genetic gap was detected north of the two geographic gaps.

Chloroplast SSR

A total of twenty-four alleles were detected in ccmp7. All individuals were successfully amplified at this locus. The allelic richness showed a significant correlation with latitude, with highest values in the north. The same trend was observed for gene diversity, but the cline was not significant. The distribution of haplotypes was well ordered with allele length increasing from north to south.

Population Structure

The haplotype distribution pattern was mirrored in the NJ tree showing two northern and southern groups. Significant IBD was found and also, alleles seemed ordered in one northern and one southern lineage. Accordingly, R_{ST} was significantly higher than G_{ST} (0.689 and 0.203, respectively, $P < 0.01$) indicating the presence of a phylogeographic structure. The level of population differentiation obtained with G'_{ST} was 0.87, which is higher than the value detected with nSSRs ($G'_{ST} = 0.33$). Overall, the genetic structure in the cpSSR was clearly different from the one detected in nSSRs, but was in line with the nSSR in that there was no relationship with the two geographical gaps.

NUCLEAR AND CHLOROPLAST SSRS

Comparison of Individual-based SGS between nSSRs and cpSSR

The SGS pattern was different between the genomes and the values of spatial autocorrelation coefficients and r were positively significant from the first to the third distance classes in the cpSSR while it was significant only in the first distance class for nSSR. Significantly negative values were reached in the sixth and seventh distance class for the cpSSR. The heterogeneity tests showed that the entire SGS was different between the two genomes. This indicates that the nuclear genome is more locally structured than the cp one, as spatial autocorrelation was found over larger distances in the cp marker.

Demographic History

Scenario 3, which corresponds to a simple simultaneous split of the four populations had by far the highest posterior probability (0.6146, 95%, CI = 0.5596–0.6697). The median values of the effective population size were 9230, 6220, 5610, 9040, 6770, and 4810, for N_A (PopA), N_B (PopB), N_C (PopC), N_D (PopD), N_1 (Pop1) and the ancestral N_2 (Pop2), respectively,. The median values of the divergence time, *t*2 and the time of population size change, *t*3 were 1760 and 4340 generations ago, respectively. If we assume a generation time of 25 years, the divergence time of the three populations would be 44,000 years ago and the time of the ancestral population size change 108,500 years ago. However, the posterior distribution pattern suggested that *t*3 is poorly estimated. The median value of the mutation rate of SSR and SNI at examined loci were estimated at $5.09 \times 10^{"4}$ and $2.38 \times 10^{"5}$, respectively. Observed values of the expected H_E, the number of alleles (A) in each population and H_E, A and F_{ST} for all possible combinations of population pairs did not differ significantly from simulated values based on parameters values drawn from the posterior distributions for scenario 3. In the PCA, the observed data (large yellow dot) is among the values obtained from the posterior distribution (large blue dots), the small dots corresponding to the prior distribution. This indicates a good fit of the posterior distribution based on scenario 3 to the data.

DISCUSSION

Genetic diversity of D. Malabaricum

In the present study we investigated genetic diversity and population structure across the range of *D. malabaricum* in the Western Ghats using both nuclear and cpSSRs. Firstly, as regeneration was poor, potentially because of overharvesting or newly generated niche competition induced by human land use, we aimed to compare genetic diversity among age classes, expecting lower genetic diversity in younger individuals. However, no difference was found between them. This pattern was also detected in a fine-scale genetic study of *D. malabaricum* in the southern part of the Western Ghats. Indeed, as trees are typically harvested after about 80 years, old adult trees are still relatively frequent. Thus, these older trees would still contribute to the reproductive success and help maintain the genetic diversity in younger cohorts. In addition, as clusters of populations were detected by the STRUCTURE analysis, sufficient gene flow between populations within clusters may occur, and it would also contribute to the maintenance of genetic diversity in individual populations. Although gene diversity and allelic richness did not show geographic patterns in nSSRs, the F_{IS} values were significantly higher in the northern populations than in the southern populations, suggesting that inbreeding is higher in the north. This pattern gives support to the classification of populations according to their perceived level of disturbance as populations classified as highly disturbed (category C) are concentrated to the north, whereas most of the least disturbed populations (category A) are found in the south.

Even in the southern part of the Western Ghats (around the Coorg population which showed significant deviation of F_{IS} from 0 in this study), a recent study on parentage and kinship analysis of *D. malabaricum* in a fragmented agro-forestry landscape found that pollen dispersal between sacred groves prevented the build-up of inbreeding at different life stages (adults, sapling, seedlings, and embryos). Data from both southern and northern populations hence suggest that the higher values of F_{IS} in the northern populations detected in this study could indeed be due to inbreeding induced by small population sizes and relative isolation at the margin of the species together with serious disturbance. This interpretation is further supported by the detection of a bottleneck in the northern most population, Yakambi (1) and by the observation of a smaller number of adult trees in this population (<20 adult trees) than in other northern marginal populations (e.g., Hittalahalli, Navangere, and Sarekoppa) (>60 individuals, Y. Tsuda and G. Ravikanth, field obs.).

Population Structure and Demographic History

Three and two genetic clusters were observed with nSSRs and cpSSR, respectively, but notably, the Palghat and Shenocottah gaps did not appear to be reproductive barriers for either genome. The distributions of the clusters were not congruent, for nuclear and cp markers and an explanation could be that population demographic history differs between the genomes, reflecting differences in inheritance and dispersal modes. The cpSSR revealed a clear geographical gradient in distribution of ordered haplotypes, which is likely informative although only one locus was examined.

Genetic clusters in the nuclear genome appear well defined over all populations except Agumbe, in which an admixture-like pattern was detected by the STRUCTURE analysis. However, the ABC analysis suggested that the most likely scenario was a simple population split into four population groups. The divergence time of the four groups was estimated to 1760 generations ago, or 44,000 years ago when we assume a generation time of 25 years for *D. malabaricum*. Estimating generation time remains difficult in species with long reproductive spans like forest trees. While *D. malabaricum* bears first fruits at

12–15 years of age it only reaches the canopy after around 25 years at which stage the trees acquire their full reproductive potential. If we assume that the generation time is around 25 years then the divergence time predates the last glacial maximum (LGM, 26,500–19,000 years before present). During the LGM the Western Ghats shifted toward colder and drier climate than today, severely affecting the distribution of rainforests. In India, rainforests spread from isolated pockets only 4000–7000 years ago. Thus, these facts together with the results of the ABC analysis suggested that the divergence of *D. malabaricum* occurred before the LGM, and that refugia were formed in several places. Our results from the STRUCTURE and ABC analysis support a model where modern populations of *D. malabaricum* were founded from one refugium in the south and two or three refugia in the north. Indeed, one such rainforest refugium has been found in southern Western Ghats in a palynological study. Also, although *D. malabaricum* has a clinal distribution, important environmental variables such as mean annual rainfall and length of dry season, remain largely constant throughout the range. In conclusion, the overall population genetic structure was therefore likely formed by historical gene flow and past climatic events rather than recent human activities and adaptation to local climate.

The finding that an instant split into four groups fit the data better than a model with three groups, where two of them merged to found a fourth, admixed population is in itself interesting. This suggests that the Agumbe population is a unit of its own, rather than an admixed population. This interpretation is supported by a recent study where ancestral polymorphism was explicitly taken into account in the case of freshwater fish. In this study ABC analysis was performed on simulated and empirical data, and it was found that the genetic structure was better explained by a population split model without admixture (a similar model to scenario 3 in the present study) than by a model with admixture, even when STRUCTURE analysis showed an admixture-like pattern. In fact, one of the issues with inferences of population demography is that ancestral shared polymorphisms are often difficult to separate from admixture or gene flow, and are extremely common in trees. Thus, the ABC approach in the present study provides demographic information for the conservation of relevant units in the target species. However, there are some caveats. As pointed out earlier the ABC analysis implemented in DIYABC assumes no gene flow among populations. The rather low level of population differentiation (overall $G'_{ST} = 0.33$) suggests that there may be some gene flow, although the clear STRUCTURE results suggest that recent gene flow may not be too important. Therefore, estimates of divergence time and effective population sizes will likely be biased downwards and upwards, respectively. However, the main results of pre-LGM divergence would not be changed even when gene flow is taken into account as ignoring gene flow would lead to underestimating the divergence time.

Additional insights on the demographic history of *D. malabaricum* can be gained by comparing nSSRs and cpSSR. Since *D. malabaricum* is insect pollinated and seeds are dispersed via birds, one would expect more genetic structure in the nuclear genome than in the cp one, which was indeed found. The Coorg population primarily belonged to the southern cluster in the STRUCTURE analysis, while it was assigned to the northern group based on cpSSR variation. This suggests that organelle capture occurred between the northern and the southern lineages in the middle of the species range by repeated hybridization and backcross events among them through pollen flow from the southern lineage. However, it should be noted that we cannot rule out the possibility that the contrasting patterns could stem from a difference in mutation rates between the cp and nuclear genomes.

Although there was some discrepancy in geographic distributions of genetic groups between nSSRs and cpSSRs groups, both genomes showed clear clusters, suggesting reproductive barriers in the species distribution range and historically limited gene flow among them. However, the Palghat and Shencottah gaps did not constitute such barriers. Regarding the seed flow, although the feeding behavior of hornbills tends to cause aggregation of individuals, they have been observed to fly over unsuitable habitats between fragments and sometimes even migrate several hundred kilometers. This ability is apparently sufficient to maintain high level of genetic connectivity between populations within the northern and southern cpSSR groups of *D. malabaricum*. Therefore, it may not come as a surprise that the 40-kilometer wide Palghat gap did not constitute a barrier to gene flow. On the other hand, while the longest pollen dispersal distance was 23.6 km, the mean pollen dispersal was 1205 m in high-density stands and 600 m in low-density stands, via beetles and thrips. Thus, pollen dispersal in this species appears more restricted than seed dispersal. The results of the SGS analysis supported this and showed wider significant spatial autocorrelation in the cpSSR than in nSSRs. Moreover, in spite of multimodalities in the STRUCTURE analysis, clear clustering was also found and several single populations were assigned to specific clusters for number of K as high as $K = 8$ and this genetic differentiation at a local scale also suggested limited pollen flow among populations.

Implications For Conservation

The geographic patterns of genetic diversity of *D. malabaricum* that was found appears to reflect the species' natural population history rather than recent human impact. However, although clear loss of genetic diversity by human activities was not suggested by this study, ongoing activities (e.g., harvesting and deforestation) and its secondary impact (niche competition in the ecosystem) as reflected in the imbalance between age classes are still serious concerns for the long-time survival of the species. Especially, evidence of inbreeding in the northern populations, which are also among the most disturbed ones, and of a recent bottleneck in the Yakambi population, may be the first signals of unsuccessful regeneration due to human activities. Our study hence suggests that conservation priority should be given to these northern populations and it is recommended to start efforts to evaluate regeneration dynamics in this region. This would entail fine-scale analysis of genetic structure among age classes. Our data also suggest the presence of long distance seed dispersal at least within the two groups detected with cpSSR. Given the importance of long-distance dispersal in *D. malabaricum*, it is desirable to maintain populations of its seed disperser, the Malabar grey hornbill. Luckily, this species has recently been listed by the IUCN as "Least concern". However, as discussed above, the distributions of haplotypes were largely different between the northern and southern groups. Therefore, more populations would be needed between the two groups to evaluate the distribution of the genetic barriers of *D. malabaricum*.

Wide-scale genetic structure studies have provided useful information on conservation units and seed zones and guidelines for restoration or plantation efforts in forest conservation and tree breeding programs of many tree species, especially economically important ones. This information provided by wide-scale genetic structure could also be relevant to the conservation of *D. malabaricum*. Based on the present study, we suggest that *D. malabaricum* should be managed as four separate units, according to the clusters found with nSSRs. Although the population structure of cpSSR was not congruent with this, priority should be given to the results of the nSSRs for two reasons; the inference is based on a larger number of independent loci, and it revealed more fine-scale diversity which might be lost if the species were to be managed as only two units. Therefore, the minimum ambition should be to conserve one population each from the four clusters as a representative for their respective genetic diversity. The status of the Agumbe population as a separate unit is not fully settled but should it be an admixed population, it still harbors genetic diversity from all the three other known lineages and thus specific conservation efforts would be justified. Although currently large-scale transplantation or restoration of individuals may not be needed in order to mitigate inbreeding over most of the species' range, the proposed conservation units would be informative for adaptive management of this natural resources. Moreover, it is required to evaluate the ecological dynamics of not only the target species but also the whole ecosystem associated to it and to design a more practical approach of ecosystem management in this biodiversity hotspot.

8

Thrissur and Kerala Forest

EVERGREEN FORESTS

SEMI-EVERGREEN FORESTS

Semi-evergreen forests (west coast semi-evergreen forests) are generally considered as a transitional stage between evergreen and moist deciduous forests. It is also found in localities where the evergreen forests are subjected to high disturbances. These forests occur between 600 to 800 m and in some places it extends up to 900 m. Animal species such as lion tailed macaque, Nilgiri langur, Nilgiri marten, small Travancore flying squirrel, brown mongoose, Malabar civet, and many birds such as the great Indian hornbill and the Bourdillon's great eared night jar occupy specific niches in these forests.

The floristic composition is an admixture of both evergreen and deciduous species in the top storey. The prominent evergreen species are Artocarpus heterophyllus, Bischofia javanica, Calophyllum elatum, Euvodia lunuankenda, Hopea ponga, Mangifera indica, Mesua ferrea and Myristica dactyloides. The deciduous floral elements include Acrocarpus fraxinifolius, Bombax ceiba, Chukrasia tabularis, Dalbergia latifolia, Grewia tiliaefolia, Lagerstroemia microcarpa, Pterospermum sp., Terminalia bellirica and Toona ciliata. The species occurring in the lower layer are the same as seen in the evergreen forests.

SOUTHERN HILL TOP TROPICAL EVERGREEN FOREST

It is an inferior variety of the typical evergreen forest, reaching to a maximum height of only 10 m.

Distribution: This type of forest abounds in the Andamans and Western Ghats. They are usually seen on the slopes and tops of hills.

Locality factors: High winds, less favourable soil and climatic conditions restrict the formation of a climax. Rainfall is usually high, over 4500 mm and humidity is high even during periods of scanty rainfall.

Floristics: Top canopy trees & Second storey trees - Artocarpus heterophyllus, Canarium strictum, Cedrela toona, Cullenia exarillata,

Dysoxylum malabaricum, Elaeocarpus seratus, Eugenia species, Holigarna beddomei, Mesua ferrea.

Bamboos :– Ochlandra travancorica

Shrubs: Pandanus spp, Strobilanthes spp

Climbers: – Calamus spp

West coast tropical evergreen Forest

These are dense evergreen forests with lofty trees of 45 m or more height. A large number of species occur mixed together. This makes the canopy extremely dense. Ferns, mosses, aroids and orchids are seen in plenty. The undergrowth consists of cane, creeping bamboo, and palms. With the increase in elevation and rainfall, the height of the forest diminishes, though it remains dense and evergreen, changing into the stunted wet sub-tropical forest.

Distribution: Enjoys a wide distribution over the Western Ghats

Locality Factors: It is seen in an altitudinal range of about 250-1200 m.The rainfall varies from 1500-5000mm.

Floristics: These forests are characteristic in having a high proportion of Mesua ferrea, Palaquium ellipticum, Cullenia exarillata and Calophyllum elatum. The absence of Hopea parviflora and Dipterocarpus indicus needs mention.

Top canopy trees: Artocarpus hirsutus, Bischofia javarnica, Canarium strictum, Calophyllum elatum, and Dysoxylum malabaricum

Second storey trees: Actinodaphne hookeri, Cinnamomum zeylanicum, Euphoria longana, Myristica beddomei, Vateria indica.

Shrubs: Leea indica, Pandanus spp, Strobilanthes spp, Rubiaceae

No grass in undistributed forest.

Climbers :– Climbers on the whole are not woody

Wet evergreen and semi-evergreen climax forests

In Kerala, wet evergreen forests are mostly confined to the windward side of the WG, where the rainfall is above 2000mm. By taking into account the distribution pattern of certain charactristic species, which reflect the climatic variations, the forests are further subdivided into eight main floristic types and three facies. All these types are classified according to low (0-800m), medium (800-1450m) and high (1400-1800m). The medium elevation forests in some places may appear at lower elevation (650 m) due to local variations in the moisture and exposure.

DECIDUOUS FORESTS

Based on the moisture regime, moist deciduous forests are divided into primary/climax or secondary moist deciduous forests. The primary moist deciduous forests generally occupy the rainfall zone of 1500 to 1800 mm, as a transition between wet evergreen and dry deciduous forests. The secondary moist deciduous forests occur within the potential area of wet evergreen formations, where the rainfall is more than 2000 mm. Although the floristic composition is almost similar in both the types, the relative dominance of certain species varies.

SECONDARY DRY DECIDUOUS FORESTS

These are inferior climax forests predominated by poorly shaped, small sized trees. Sandal is also seen in such forests.

Distribution: They are seen distributed in dry deciduous forests and intruding into the drier parts of moist deciduous forests.

Locality Factors: The soil surface is hard and impervious due to exposure and trampling effected by heavy grazing, fuel and timber collection.

Floristic: Top canopy trees – Bombax ceiba, Grewia tiliaefolia, Schleichera oleosa, Tectona grandis

Second storey – Feronia limonia, Santalum album Shrubs – Dodonaea viscosa, Lantana camara

Southern Dry Deciduous Forests

The sub group differs from the dry teak forest species-wise, though typical plants like Boswellia are conspicuous. Heavy grazing invigorates growth of thorny species. Bamboo is mostly absent and of poor quality, if present. Climbers are rarely seen.

Distribution: It occurs throughout peninsular India, especially in drier localities.

Locality factors: The rainfall varies from 875 mm –1125mm on dry sites and soils. The shallow soiled, well-drained hillsides and the undulating grounds have identical forests, making it difficult to establish the relation of site and climate to the forest in situ.

Floristic: Diospyros tomentosa, Chloroxylon swietenia, Hardwickia binata, Boswellia serrata

Primary moist deciduous forests (Lagerstroemia microcarpa – Tectona grandis – Dillenia pentagyna type – LTD)

Primary deciduous forests are found in isolated patches between the Anamalai and Wayanad plateaus. Denser part of this type is the form of woodland and savanna woodland. Dillenia pentagyna and Tabernaemontana heyneana are characteristic species of this type. Lagerstroemia microcarpa and Tectona grandis, together with other species such as Anogeissus latifolia, Dalbergia paniculata, Pterocarpus marsupium, Terminalia paniculata, Hymenodictyon excelsum, Haldina cordifolia are common.

Secondary Moist Deciduous Forests:

In kerala secondary forests cover larger areas than the primary type, mostly in the form of dense forests and woodland to savanna woodland. Especially on the steep slopes, they are found as tree savanna. Floristically, there are similar to primary moist deciduous however, some deciduous species like Dillenia pentagyna, Tabernaemontana heyneana, Strychnos nux-vomica, and Xylia xylocarpa, are relatively more common than in the primary forests. Tectona grandis, which is extensively planted, has also been found mixed with other species in dense formations. In the dense forests often there is dominance of evergreen species like Ixora brachiata, Olea dioica, Persea macrantha, Dimocarpus longan, Flacourtia montana etc.

Dry Deciduous Forests

With in the given rainfall regime, dry deciduous forests in Kerala State are rare. They are confined to northern slope of Anamalai in Chinnar Wild life Sanctuary, eastern part of Mannarkad Division, and South Wayanad Wildlife Sanctuary where the rainfall is less than 1200 mm. The physiognomic structure of these dry deciduous forests is highly variable, due to impoverishment of soil, especially on steep slopes, and also due to anthropogenic pressures including fire and grazing. Three types of dry deciduous forests have been recognized.

Albizia amara –Acacia spp. Gyrocarpus Asiaticus Type (AAG)

This type is found only in Chinnar Wild life Sanctuary, up to 650 m. On the lower slopes, Acacia chundra and A.leucophloea are characteristic species, particularly in the scrub woodland and thickets. Albizia amara, Erythroxylum monogynum, Dichrostachys cinerea and Chloroxylon swietenia, and Hardwickia binata are the other common species of this type.

On the slopes, especially on skeletal soils, tree savannas are the prominent formations. In such habitat Gyrocarpus asiaticus, with metallic-coloured bark, is the characteristic species, along with other slope-loving species, like Cochlospermum religiosum, Givotia rottleriformis, Sterculia urens and Commiphora caudate. Anogeissus latifolia – Pterocarpus marsupium – Terminalia spp. type (APT) This type is found above 600 m in Mannarkad Division (northern part) and Chinnar WLS. As it is generally found on slopes, physiognomy varies from savanna woodland to tree savanna. Apart from the species mentioned in this type Dalbergia paniculata, D.latifolia, Emblica officinalis, Kydia calycina and Grewia tiliifolia are also common.

Anogeissus latifolia – Tectona grandis – Terminalia spp.type (ATT)

This type is found only in the South Wayanad WLS. It is generally represented by dense forest and woodland to savanna woodland. Compared to adjacent primary moist deciduous forests, here the species like Dillenia pentagyna, Alstonia scholaris, Callicarpa tomentosa disappear and the species mentioned in the type become dominant. Other common species include Diospyros melanoxylon, Madhuca latifolia, Emblica officinalis, Lagerstroemia parviflora Careya arborea etc. In some poorly drained low-lying areas Shorea roxburghii become conspicuous.

GRASSLANDS

In Kerala grasslands are generally found above 1500 m. The grasslands, which are also called as 'shrub-savanna' are characterised by herbaceous and shrubby species mixed with grasses.

The grasslands below 1800 m that are adjacecnt to medium or high elevation evergreen forests, are often found with sparse trees, represented by Wendlandia thrysoidea, Glochidion spp. Terminalia chebula, Emblica officinalis, Careya arborea, Briedelia crenulata; in some places a dwarf palm. Phoenix is found in patches. At this elevation range, grasses are tall, and reach the height up to 1.5 m. They are commonly represented by Androprogon lividus, Arundinella purpurea, Agrostis peninsularis, Chrysopogon zeylanicus, Eulalia phaeothrix, Sehima nervosum, Heteropogon contortus, Eulalia sp, Themeda sp, Ischaemum indicum, and Tripogon bromoides. In cattle grazed and frequently burnt areas, unpalatable Cymbopogon flexuous and Pteridium, a fern are frequent.

The grasses in this zone are mixed with other herbs like Crotalaria, Desmodium, Hypericum, Knoxia, Leucas, Lobelia, Osbeckia etc. Phlebophyllum kunthianum, a monocarpic shrub species, often dominates the grass land landscape.

At above 1800 m, especially in the Anamalai region (Eravikulam and Munnar) grasslands are more specialised. During the colder months, the minimum temperature often goes below zero degree centigrade. In this zone grass layer is less than 1m and is represented by Andropogon foulkesii, Anthistiria ciliata, Arundinella spp., Arundinaria villosa, Bothriochloa pertusa, Chrysopogon orientalis, Cymbopogon spp.,Eragrostis nigra, Eulalia spp., Heteropogon contortus , Isachne spp., Themeda spp., Tripogon bromoides and Zenkeria elegans.

Among Shrubby elements Berberis tinctoria, Gaultheria frangrantissima, Hypericum mysorense, Lobelia excelsa, Oldenlandia stylosa, Osbeckia wightianum, Pteridium aquiilnum, Rubus fairholmianus, Phlebophyllum kunthianus are particularly frequent. Rhododendron arboreum var. nilagiricum in the form of small tree is also sporadically seen in grasslands.

The common herbaceous elements among grasses include Anaphalis spp., Campanula fulgens, Cassia spp., Crotalaria notonii, Cyanotis spp.,Indigofera pedicellata, Justicia simplex,Knoxia mollis, Leucas suffruticosa, Lilium neilgherrense, Oldenlandia articularis Polygala sibirica, Striga asiatica, Viola patrinii,and Wahlenbergia gracilis. In the swampy pockets Commelina spp., Centella asiatica, Drosera peltata, Fimbristylis uliginosa etc are common. .

MANGROVES

Mangroves are wetland ecosystems formed by the assemblage of specialized plants and animals adapted to semi saline swamps along coasts.

Mangrove forests of Kerala are highly localized, but the species diversity of these mangroves and its associates are comparatively rich. It is confined to the upper reaches of estuaries, lagoons, backwaters and creeks. In Kerala mangroves are distributed in all the districts except Idukki, Pathanamthitta, Palakkad and Wayanad. Maximum extent is reported from Kannur district. The total extent of mangrove forests in the state is estimated to be less than 50km^2. Mangroves play an important role in the economy of coastal people through various ways. Mangroves provide excellent habitat for migratory birds, serve as breeding ground for many species of fishes and prawns helps in controlling pollution, rutting of husks etc.

The important mangrove plants are:

Acanthus cillicifolius, Acrostichum aurem, Aegiceras corniculatum, Avicennia officinalis, A, rina, Azima tetracantha, Bruguiera gymnorrhiza, B. cylindrica, B sexangula, Excoecaria agallocha, E indica, Kandelia candel, Rhizophora apiculate, R mucronata, Sonneratia caseolaris, Calophyllum etc. Some of these species that disappeared from the Kerala coast are Azima tetracantha an Ceriops tagal, Heritiera littoralis and Flagellaria indica have discourteous distribution. Calamus rotang and Syzygium travancoricum are some of the rare and endangered species found in the mangroves.

The major threats to the mangrove forests are land reclamation for urbanization, intensive aquaculture felling of mangrove trees for fuel and fodder, unsustainable land use, ambiguity in ownership etc.

CULLENIA EXARILLATA- MESUA FERREA- PALAQIUM ELLIPTICUM – GLUTA TRAVANCORICA TYPE (CMPG)

This type is confined to south of the Ariankavu Pass, between 8°20'N and 8° 50' N. It is mainly defined by the altitudinal preference of Cullenia exarillata. This speceies is commonly found between 700 and 1400 m and seldom descends to 550 m in some moist valleys. Its latitudinal distribution goes up to 11° 55'N (Wayanad plateau.) Mesua ferrea and Palaqium ellipticum are widely distributed in both the low and medium elevation forests, however they tend to become more dominant at medium elevation. Gluta travancorica, a common canopy species is endemic to the south of the Ariankavu pass. The association of this species with the other three species (CMP) makes them a distinct medium elevation type.

There are certain species, which are either exclusive to the CPMG type or rarely found in the other medium elevation type north of Ariankavu pass. They are represented by Calophyllum austroindicum, Garcinia rubro-echinata, Garcinia imbertii,Garcinia travancorica, Diospyros barberi, Atuna travancorica, Nageia wallichiana at canopy and subcanopy level; Memecylon subramanii, Popowia beddomeana, Memecylon gracile, Octotropis travancorica, Diotacanthus grandis, Goniothalamus rhynchantherus and Vernonia travancorica at undergrowth level.

Nageia wallichiana (Podocarpus) a sole indigenous Gymnosperm tree species in South India. In the WG it is found in the Agasthyamalai region (Periyar plateau) and in Anamalai region. In the Agasthyamalai region it is more common towards the eastern side as a canopy tree it forms with CMPG type, at above 1000 m. In the other two regions, they are rare in the CMP type of forests.

The species, which are also characteristic in this facies include Elaeocarpus venustus, Eugenia floccosa, Aglaia bourdilloni, Actinodaphne campanulata and Syzygium microphyllum. Bentinckia codapanna, an endemic palm is often found along the margins of the facies near the cliffs.

CULLENIA EXARILLATA –MESUA FERREA – PALAQUIUM ELLIPTICUM TYPE (CMP)

This type occupies the WG of Kerala between the Ariankavu Pass and the Brahmagiri Ghat in Wayanad. The lower limit of the type on the western side of the Ghats locally varies from 550 to 750 m. The bioclimatic range of this type is generally similar to CMPG type except in the length of the dry season, which varies from 2 to 4 months.

In the Elaimalai region, the entire CMP type has been converted into cardamom plantations. In these plantations, cardamom has been grown as undergrowth, while keeping the original canopy intact. In the dense forests, apart from the type species,other common canopy tree or emergent species are Diospyros sylvatica, Drypetes elata, Cinnamomum keralense, Syzygium gardneri, Dimocarpus longan, Aglaia lawii, Litsea oleoides and all these are widely distributed.

In the second and third ensembles that are also broadly distributed include Agrostistachys meeboldii, Symphilia mallotiformis, Tricalysia apiocarpa, Myristica dactyloides. Homalium travancoricum and Diospyros paniculata. However, some species like Drypetes venusta, Semecarpus travancorica, Diospyros nilagirica, Litsea bourdillonii, Litsea keralana, Bhesa indica, Aglaia tomentosa , are found up to the Palghat Gap. Beyond the Gap, these species either disappear or become rare. In the fourth ensemble Ardisia pauciflora,Goniothalamus wightiana, Tabernaemontana gamblei, Psychotria anamalayana, Lasianthus jackianus are common.

Dipterocarpus Indicus – Dipterocarpus Bourdillonii - Strombosia Ceylanica Type (DDS)

This type covers a wide area in the WG between Ariankavu Pass and northern border of the Palaghat Gap (9°n and11^{o}N). Its general bioclimatic conditions are similar to that of the DKS type, except for the mean temperature of the coldest month, which is higher than 20°C

Like in the previous type, Dipterocarpus indicus and Strombosia ceylanica are also characteristic species in this area. However, Kingiodendron pinnatum, which is dominant in the earlier types, has been replaced by gigantic

Dipterocarpus bourdillonii. The distribution of both the Dipterocarpus are patchy due to over exploitation in the past. Dipterocarpus bourdillonii, which generally occurs at lower limits(<450 m), is the most affected species. Currently it is encountered only along streams and in some inaccessible areas.

Bulk of the DDS type is in the form of distributed forests. These forests were logged for several decades for hardwoods and softwoods. Due to this repeated logging several climax species like Vateria indica, Palaquium ellipticum, Calophyllum polyanthum, Otonephelium stipulaceum, Chrysophyllum roxburghii, Dipterocarpus indicus, Semecarpus auriculata, Poeciloneuron indicum, have become less frequent. However, species with wider ecological aptitude viz. Polyalthia fragrans, Pterygota alata, Artocarpus gomezianus, Antiaris toxicaria and Bombax ceiba have become common and taken over the canopy.

DDS type is also dotted with swamps especially in Ranni (between Plapally and Erumeli) and Kannur Divisions. Some of these swamps are dominated by Humboldtia vahliana with looping stilt roots along with Myristicaceae members viz. Knema attenuata, Myristica dactyloides and Gymnacranthera canarica

DRY TEAK FOREST

This sub-type occurs in areas having a rainfall of 900 mm-1300mm. It is characterized by shallow, porous or clayey soils and heavy grazing combined with frequent fires.

Floristics: Top canopy trees – Tectona grandis, Anogeissus latifolia, Pterocarpus marsupium, Boswellia serrata,Terminalia tomentosa, T. paniculata, T bellirica Second storey trees: – Chloroxylon swietenia, Diospyros montana, Hardwickia binata, Cassia fistula, Wrightia tinctoria

High elevation type

The high elevation forests are generally confined to altitude between 1400 and 1800 m, where the mean temperature of the coldest month varies 14-16°C and length of the dry period ranges from 2 to 3 months. Rainfall in the area varies between 3000 to 5000 mm. Structurally, the forests are stunted, with two ensembles, and canopy seldom exceeds 15 m. Two floristic types have been identified in these high elevation forests.

Bhesa indica – Gomphandra coriacea – Litea spp. type (BGL)

In the Western Ghats of Kerala , this type is found between the Ariyankavu Pass and Palghat gap, between 1400 and 1800 m. At this elevation range, several species of the lower elevations disappear or become very rare viz Cullenia exarillata, Palaquim ellipticum, Diospyros spp and Agrostistachys meeboldii. The family Annonaceae which is dominant at lower strata in both low and medium elevations, disappears completely at high elevation. However, some species that are less important at lower elevations become significant at higher elevation range. Gomphandra coriacea, vicariant of Gomphandra tetrandra of low elevation forest, become conspicuous at lower strata. Lauraceae, which tends to become common with the increase of elevation, manifest its highest diversity. Several other species like Schefflera capitata, Mastixia arborea, Archidendron clyparia, Hydnocarpus alpina, Cocculus laurifolius, Acronychia pedunculata, Isonandra spp. Meliosma spp., Symplocos spp are also common in this type.

Schefflera spp. Meliosma Arnottiana - Gordonia obtuse Type (SMG)

This type found in north of the Palaghat gap (between 1400 and 1800 m) as a transition between CMP of medium elevation and LSM of montane type. Compare to its counter part (BGL) type in the south of the Palaghat gap, SMG type requires 3 to 6 months dry period.

Even though the species mentioned in this type is found through out the medium elevation and above, they reach their optimum presence in the above mentioned region. Among Araliaceae Schefflera capitata, S. micrantha, S.racemosa. S.wallichiana are common. Lauraceae (Litsea, Cinnamomum, Alseodaphne, Neolitsea) and Myrtaceae (Eugenia, Syzygium, Rhodomyrtus) become conscpious from this type and towards montane type.

LATERITIC SEMI EVERGREEN FOREST

Distribution: These are forests, which come up in the latteric soils and characterized by the presence of Xylia xylocarpa.

Floristics: Top Canopy trees – Xylia xylocarpa, Pterocarpus marsupium, Anogeissus latifolia, Grewia tiliifolia,

Secondary storey trees – Briedelia retusa, Strychnos nuxvomica, Calycopteris floribunda

Shrubs – Adhatoda vasica

LITTORAL FOREST

This type of forest occurs along the coast having a fair width of sandy beach. The most important species is the tall evergreen, light-foliaged casuarina. In the absence of casuarina, smaller evergreen and deciduous trees form the dominant canopy.

Locality Factors: The habitat exhibits pecularities, which needs mention. The sea sand has adequate lime content, but little nitrogen and mineral nutrients. Though the beach sand is coarse and porous, a high water table of brackish water is always maintained due to the proximity to the water body. The high insolation levels and sand-laden winds render the area suitable only to xerophytic plants. The mean annual temperature ranges from 26-29 degree Celsius. The rainfall received is around 5000 mm.

Floristics: Species composition varies in different regions

LOW ELEVATION TYPES

The low elevation types are typically 'Dipterocarp' Forests. Structurally, they are all dense forests with four structural ensembles (sense Oldeman, 1974) and an emergent layer. Canopy height often reaches 35-45 m. Floristically, low elevation forests are grouped into three main types

Dipterocarpus indicus: – Kingiodendron pinnatum – Strombosia ceylanica type (DKS)

This type is confined to south of the Ariankavu Pass (8° 20'N to 9° 00'N), where the length of the dry season varies from 2 to 3 months. Like in the other two low elevation evergreen forest types, Dipterocarpus indicus is a characteristic canopy species. Kingiodendron pinnatum, yet another important canopy tree in the type, shows a peculiar distribution pattern. It is very rare in the low elevation wet evergreen type between the Ariyankavu Pass and the Palaghat Gap, however, becomes prominent again north of the Palaghat Gap. Strombosia ceylanica has a wide distribution throughout the WG, but is more common among the canopy trees of low elevation types, south of the Palaghat Gap.

The denser part of the DKS type is represented by two facies, which are found above 8 ° 40'N. Below, that, the DKS type is mostly in the form of fragments or in a degraded condition. Some of the big fragements are close to climax forests with the presence of Vateria indica, Kingiodendron pinnatum, Hopea parviflora, Mesua ferrea, and Strombosia ceylanica. Dipterocarpus indicus are rare in these fragments.

The two facies that have been identified under DKS type, one is characterized by the local abundance of otherwise two geographically rare species (Hopea racophloea and Humboldtia decurrens),and the other corresponds to a particular ecosystem adapted to water logged areas.

(i) Hopea racophloea – Humboldtia Decurrens Facies

This facies is confined to humid valleys between Kallar and Shendurni Rivers , west of Agastya malai. Although, the distribution of Hopea racophloea and Humboldtia decurrens goes beyond the Ariankavu Pass, their presence in DKS type is more conscipious than in any other regions. Hopea racophloea with exfoliating bark is a canopy tree. Its regeneration is always gregarious. Humboldtia decurrens, a cauliflioru s tree with large winged pinnate leaves, is prominent in the third structural ensemble of the forest. Other common canopy trees are Vateria Indica, Artocarpus gomezianus, Otonephelium stipulaceum, Holigarna nigra, Cynometra sp. and ficus beddomei. Poeciloneuron indicum has been found in patches. semecarpus auriculata and semecarpus travancorica are common towards the lower limit of the type. Among the species of lower ensembles Diospyros paniculata, Fahrenheitia zeylanica, Diospyros humilis, Hydnocarpus macrocarpa, Knema attenuata, Cynometra bedlomei are important.

Medium elevation types

Medium elevation forests are structurally very similar to low elevation ones especially at the lower limits, they are tall (canopy 35-45 m) with four structural ensembles. Towards the upper limit, the forests are stunted with two or three ensembles (canopy < 18 m). They differ floristically from low elevation types due to disappearance of species like Dipterocarpus spp., Kingiodendron pinnatum etc. At this elevation range, the relative abundance of certain species like Strombosia ceylanica, Vateria indica, Diospyros bourdillonii etc. have also become less. Two main types and one facies have been recognized at medium elevation

Moist Teak Bearing Forest

Here, the teak found is usually of second and third quality. The overwood mostly comprises of deciduous species.

Floristics: Top Canopy Trees - Terminalia Tomentosa, Dalbergia Latifolia, Bombax Ceiba, Lannea Coromandelica, Albizia Lebbeck

Second storey trees – Xylia xylocarpa, Wrightia tinctoria, Mallotus philippensis

II a – Bambusa bambos, Dendrocalamus strictus

Shrubs – Helicteres isora, Desmodium spp

Montane evergreen forests (Laitsea spp, - Microtropis spp – LSM):

Montane evergreen forests (sholas) are confined to high altitude plateaus, in the regions of Anamalai (Eravikulam and Munnar) ,Vavul malai (North Nilambur) and upper reaches of New Amarambalam RF (South Nilambur) , adjoining the Nilgiri plateau. These plateaus generally come under high rainfall (>5000mm) zone and the temperature (mean temperature of the coldest month) is less than 13.5°C. Here the evergreen forests are found amidst the grasslands in valleys, depressions and on convex mounts. The tree layer in shoals comprises of short boled evergreen species with canopy height up to 15 m and bark covered with lichens, orchids and moss.

Lauraceae and Myrtaceae families characterize these forests. Litsea spp. of the former family and Syzygium spp. of the latter are dominant, along with Microtropis spp (Celastraceae). Other common species in the shola include Symplocos pendula, Elaeocarpus recurvatus, Michelia nilagirica Actinodaphne bourdillonii, generally dominated by subtropical elements like Berberis tinctoria, Daphniphyllum neilgherrense, Photonia notoniana, Rapanea spp., Rhododendron nilagiricum, Rhodomyrtus tomentosa, Symplocos spp., Turpinia cochinchinensis.

MYRISTICA SWAMP FOREST

They are fairly dense evergreen forests reaching a height of 15-30 m , with clean slender boles. The undergrowth includes mainly aroids and scitaminae. Distribution: Restricted to valleys in the tropical evergreen forest of Travancore

Locality Factors: It occurs as fringe forest on slow moving streams. The soil is sandy alluvium with high humus content.

The soil layer remains inundated from June to January'

Floristics: Top canopy trees & second storey trees- Myristica magnifica , Myristica malabarica, Lagerstroemia

speciosa, Laphopetalum wightianum, Carallia brachiata

Shrubs – Pandanus sp

IV a – Cyperaceae, Scitaminae

Climbers – Calamus sp

NILGIRI SUB-TROPICAL HILL FORESTS

These forests resemble the tropical rain forest except for the stunted growth of trees. They are less luxuriant and the trees have shapeless boles, often festooned with epiphytes. Strobilanthes usually forms dense undergrowth.

Distribution: They are seen in the Nilgiris, Anamalai, and Palani Hills of Tamilnadu and Kerala.

Locality Factors: This sub group occurs at altitudes between 1000m and 1700m on the South Indian hills. The mean annual temperature ranges from 17-22 degree Celsius. The rainfall recorded is high and varies from 1500-6600mm. The number of rainy days amounts to a Maximum of 132.

Floristics: Top Canopy trees and second or secondary storey trees – Calophyllum elatum, Actinodaphne hookeri, Canthium dicoccum, Ficus arnottiana, Persia macrantha

Shrubs – Strobilanthes sp

Climbers – Calamus sp

SOUTH INDIAN SUB TROPICAL HILL SAVANNAH

An open savannah forest with tall, coarse grass reaching a height of 2-3 m. Scattered trees of deciduous nature are also seen.

Distribution: Found in the Nilgiri and Palani Hills of Tamil nadu and Kerala

Locality factors: The rainfall varies from 1500 mm upwards and is evenly distributed.

Floristics: Top canopy trees- Dalbergia latifolia, Anogeissus latifolia.

Second storey trees – Olea dioica

Shurbs- Phoenix humilis

SOUTHERN MOIST MIXED DECIDUOUS FOREST

Top Canopy trees and Second storey trees – Terminalia paniculata, Terminalia tomentosa, Careya arborea, Phyllanthus emblica, Dillenia pentagyna, Sterculia villosa, Albizia odoratissima, Cassia fistula, Gmelina arborea, Saccharum spontaneum (Grass species)

Southern Montane Wet Grasslands They are seen over large areas on the rolling downlands. The highest parts of the forest and the forest in the depressions are subject to annual frost. Fires are frequent and grazing is heavy.

Floristics: Cymbopogon flexnosus. Eragrostis nigra, Themeda cymbaria

Southern Montane Wet Temperate Forest

They are luxuriant evergreen forests with closed canopy. The trees attain considerable girth, but are short boled and branchy. The leaves exhibit a varying range of colours, which is a distinct feature of this forests. The wet nature of the forest results in abundance of moss, ferns and other epiphytes. The canopy differentiation is not discernible.

Distribution & Locality factors: It occurs in Tamilnadu and Kerala on the Nilgiris, Anamalia, Palani and Thirunelveli hills from about 1500 m upwards. It is also found in patches of Shola of the sheltered plains. The mean annual temperature ranges from 14-17 degree Celsius. The mean annual rainfall varies from 1300-6000 mm. The soil is reddish or yellowish clay, topped by varying depths of soil rich in humus.

Floristics: Ternstroemia gymnanthera Syzygium, S.arnottianum, S.tamilnadensus, calophyllifolium, E. arnottiana, E. Montana, Rhododendron nilagiricum, Elaeocarpus spp.

SOUTHERN SECONDARY MOIST MIXED DECIDUOUS FOREST

Floristics: Top canopy trees – Terminalia paniculata, Mangifera indica, Dalbergia latifolia, Lagerstroemial anceolata, Alstonia scholaris, Xylia xylocarpa.

Second storey trees – Olea dioica, Careya arborea, Phyllanthus emblica, Callicarpa tomentose Phylanthus emblica, a Bamboo absent

Shurbs - Clerodendrum Viscosum, Helicteres isora, Glycosmis pentaphylla

Climbers – Calycopteris floribunda, Acacia chesia

Tropical riparian fringing forest (riparian forest)

The forest type is characterized by a few evergreen and semi-evergreen species restricted on the sides of streams forming a narrow fringe. In the Sanctuary the forest type is restricted mostly along the sides of the Pambar and Chinnar rivers.

The dominant species are Terminalia arjuna, Hopea parviflora, Bischofia javanica, Mangifera indica, Drypetes roxburghii, Vitex leucoxylon, Pongamia pinnata, Garcinia gummi-gutta, Mallotus stenanthus, Calophyllum calaba, Entada rheedei, Lepisanthes tetraphylla, Syzygium cumini, Schefflera racemosa, Homonoia riparia, Vitex altissima, Salix tetrasperma, Gnetum ula etc

VERY MOIST TEAK FOREST

In this type, teak forms only 10 percent of the overwood. Most of these forests are sub climaxes in semi evergreen forest and are of secondary orgin.

Floristics: Top Canopy trees – Lagerstroemia lanceolata, Grewia tiliifolia

Second Storey trees – Dillenia pentagyna, Kydia calycina

Bamboos – Bambussa bambos

Shrubs – Clerodendrum viscosum, Glycosmis pentaphylla, Lantana camara

Climbers – Spathalobus roxburghii, Acacia pennata.

West coast Semi-Evergreen forest

Being intermediate between the tropical evergreen and moist deciduous forest, these are difficult to define except describing in comparative terms. They usually include patches of both these types mixed into a mosaic. It forms a closed high forest, the dominant trees sometimes will grow very large. Buttressed stems continue to frequent both in evergreen and moist deciduous forests. The general canopy is less dense than in true evergreen and the evergreen undergrowth rather copious; climbers tending to very heavy. Bamboo is usually found well distributed in the evergreen forest. Epiphytes are abundant, including many ferns and orchids.

Locality factors: Rainfall varies from 2000-2500 mm on the plains

Floristics: Top canopy trees – Terminalia tomentosa, Dalbergia latifolia,Haldina cordifolia, Xylia xylocarpa, Artocarpus hirsutus, Hopea parviflora, Mesua lerrea,

Second storey trees- Hydnocarpus pentandra, Bischofia javanica, Mallotus philippensis, Kydia calycina

Bamboos- Ochlandra sp, Bambusa bambos

9

Geology, Rock and Soil of the Western Ghats in Thrissur (Kerala)

GEOLOGY

Mountain ranges in Deccan Plateau corroded edge are the Western Ghats making it great place to explore. According to geologic evidences such mountains were developed while Gondwana break-up of supercontinent held approximately 150 million years ago itself. Researches done by the geologists and their evidences have an indication that India's west coast was formed and developed nearly 100 to 80 mya years ago once Madagascar had broken. India's western coast could literally appear as an unexpected cliff approximately 1,000 m (3,300 ft) in elevation once break-up had happened. These hills have ample basalt rock found predominantly whose thickness of 3 km (2 mi) is common. More rocks that are abundantly found in such hills include charnockites; khondalites; granite gneiss; leptynites, metamorphic gneisses and crystalline limestone detached occurrences besides iron ores anorthosites and dolerites. Southern hills have abundant residual laterite and bauxite ores here.

GEOLOGICAL FORMATIONS IN KERALA

A formation or geological formation is the fundamental unit of lithostratigraphy. A formation consists of a certain number of rock strata that have a comparable lithology, facies or other similar properties. Formations are not defined on the thickness of the rock strata they consist of and the thickness of different formations can therefore vary widely. Three main geological formations are recognized in Kerala namely, the Archaeans (oldest rocks), the Warkalli beds of Tertiary Age (Upper Miocene to Pliocene) and the recent deposits (quarternary). These have North-South alignment.

Crystalline Rocks

(i) Dharwar formation: These occur in the Malabar area only. They are represented by garnetiferous-ferruginous quartzite, mica, talc, schist etc. and are found exposed in South-East Wayanad and North West of Gudalur.

(ii) Champion gneiss: They are seen in South and South-East Wayanad and have gold bearing veins. Rocks appear to be of post peninsular age.

(iii) Peninsular gneiss: This is one of the most widespread rock types found in Kerala. The important minerals that go to make up the rocks are quartz, Feldspars, biotite and garnite. In Cochin area, they form the most extensive rocks. The types present are biotite and hornblende gneiss. In Thiruvananthapuram area, the gneiss belongs to the peninsular suite and is made of quarts, orthoclase, mica and hornblende. Charnockite and leptynites are the most common gneisses in this area.

(iv) Charnockite: A good portion of the Western Ghats is made up of this rock. In the Travancore area the rocks are well foliated and show intrusive relationship with peninsular gneiss. They are highly garnetiferous as compared to the charnokites of North Kerala, where garnet is absent.

(v) Closepet granite: Archaean intrusions of post charnockite age are found in the Malabar region. The two intrusions of biotite granite found in Kalpetta hills and Sultan's Battery, have strong resemblance to the Dornegneiss of Hazaribagh.

Precambrian System

(i) Basic dykes: These rocks are fresh but fractured and mylonitised. They approach dolerite in composition and are found to occur in South Malabar area. The basic dykes of Cochin area are fine to medium grained and free form olivine. The more coarse grained crystalline phases are represented by gabbros. Several exposures of gabbros are found in Cochin area.

(ii) Residual laterites: A narrow zone of lateritised rock exists to the west of crystalline rocks that constitute the eastern boundary of the State. The rock exposed on the surface in this zone is a type of laterite, which exhibits characteristics different from those of the laterite, which caps the Varkalai formation. The laterite preserves the structure of the parent rock and is less compact. Below the laterite layer is the kaolin layer, the depth of which to the undecomposed rock shows gradation.

(iii) The warkalli formation: This represents the most conspicuous sedimentary bed occurring in Varkala. They are best exposed at Varkala in the cliffs near the seashore. They consist of clayey sandstone, white and variegated clay and carbonaceous clay containing thin lenses of lignite. Most of these areas are lateritised.

(iv) Recent deposits: They are mainly developed in Northern parts of Quilon and are made up of sand and silt. The lacustrine deposits of the backwater tracts of Kerala, the mud banks of the coast of Alleppey and the marine beach deposits all along the sea shore of Kerala come under the group. From the economic point of view, the area is important as it contains valuable mineral sands.

ANGADIPURAM LATERITE

Angadipuram Laterite is a National Geological Monument identified in Angadipuram town in Malappuram district in the southern Indian state of Kerala, India. The special significance of Angadipuram to laterites is that it was here that Dr. Francis Buchanan-Hamilton, a professional surgeon, gave the first account of this rock type, in his report of 1807, as "indurated clay", ideally suited for building construction. This formation falls outside the general classification of rocks namely, the igneous, metamorphic, or sedimentary rocks but is an exclusively "sedimentary residual product". It has a generally pitted and porous appearance. The name laterite was first coined in India, by Buchanan and its etymology is traced to the Latin word "letritis" that means bricks. This exceptional formation is found above parent rock types of various composition namely, charnockite, leptynite, anorthositeand gabbro in Kerala. It is found over basalt in the states of Goa, Maharashtra and in some regions of Karnataka. In Gujarat inwestern India, impressive formations of laterite are found over granite, shale and sandstone.

Apart from its use as bricks in building construction, it has other substantial economic value, since it has been established that laterites are closely juxtaposed with aluminium ore (bauxite), iron ore and nickel ore mineral deposits in many parts of Kerala.

The GSI has erected a monument at Angadipuram where the laterite formations were first identified, as one of the 26 monuments declared as National Geological Monuments, on the occasion of the "International Conference on Laterization" held in 1979.

Angadipuram is also well known as a pilgrimage centre for its famous temples, the Thirumandhamkunnu temple and the Tali temple.

GEOGRAPHY

Angadipuram is situated in the Malappuram district, which lies in northern Kerala, and is bounded on the north by Wayanad and Kozhikkode districts, on the northeast by Tamil Nadu, on the southeast and south by Palakkad District, on the southwest by Thrissur District, on the west by the Arabian Sea, and on the northwest by Kozhikode District . The geographical distribution of laterite is not limited to Angadipuram in Malappuram district but it is also found in the midland regions and highlands of Kerala. Its occurrence extends to Aleppey, Quilon, Thiruvananthapuram,Kottayam, Trichur and Cannanore districts.

Overall, in the landform of Kerala which has seven landscape ecological zones, laterites account for a major share of 50%. This land form comprises lateritic mesa, mounds, slopes and ridges. This dominant laterite setting is delimited between altitudes 50 metres (160 ft) and 150 metres (490 ft) (though found up to elevation 2,000 metres (6,600 ft) and extends from the northern end to the southern tip of the state. Further, as the topography changes a few kilometers from the sea to the east, there are numerous valleys called *elas* where patches of paddy fields,coconut and arecanut groves are seen. Incidence of laterite in other parts of India is reported in the states of Karnataka Maharashtra and Gujarat.

Apart from India, its global occurrence in the form of vast deposits of lateritic bauxites with rich production is reported in Australia, Brazil, Guinea, Guyana, Suriname andVenezuela.

Climate

The intense southwest Monsoon rainfall in Kerala (average annual rainfall is 3,107 mm) coupled with high temperatures (mean annual temperatures range from 25 to 27.5 °C in the coastal lowlands to 20 to 22.5 °C in the eastern highlands) and lush vegetation (belongs to the Malabar Coast moist forests of a tropical moist broadleaf forest ecoregion of southwestern India) has accentuated the chemical processes over the base rocks, which has resulted in the formation of laterites. In view of these conditions, the laterization process, which results in formation of laterites, is called the "Tropical disease of rocks".

STRUCTURE

Laterite is a residual product created by the natural process of rocks weathering in the hot humid climatic conditions and interaction with water, oxygen and carbon dioxide. In simple terms, it is a soil formation linked to the parent rock material that has evolved because of various powers of nature in the same manner as other types of soils such as alluvial soil, regular soil and red soil. It is also inferred that paleoclimate dating back several million years has been a causative factor in laterite formation. The residue usually consists of enrichediron, aluminum and titanium oxides in varying proportions. The residue is pitted and porous in appearance. Buchanan, who discovered this formation in Kerala, in his report of 1807 observed:

It is diffused in immense masses, without any appearance of stratification and is placed over the granite that forms the basis of Malayala. It is full of cavities and pores and contains a large quantity of iron in the form of red and yellow ochres. In the mass, while excluded from air, it is so soft, that any iron instrument readily cuts it, and it is dug up in square masses with a pick-axe, and immediately cut into the shape wanted with a trowel or a large knife. It, very soon after, becomes as hard as brick, and resists air and water much better than any bricks that I have seen in India...

In the midland region of Kerala where lateritic soil is predominant, laterites form a residual deposit due to weathering of either crystalline or sedimentary rocks with thickness varying from 5–8 metres (16–26 ft). They also form plateaus. These laterite plateaus are attributed to various phases of uplift of the land in terraced formations in this region. However, in Malappuram, Kozhikode and Kannur districts in the plateau region, laterites are of greater thickness. It is also noted that top layer of laterite, over the crystalline rocks, is very compact. It is also reported by GSI that in Kerala:

Quartz veins, joints and fractures can be traced from the top to the bottom of the laterite profile. The laterite profile over pyroxene granulites, metaultramafites and gneisses are characterised by relict foliation that conforms to those of the subjacent rocks which indicate the in situ nature of the laterite. Porous and spongy texture is discernible in laterites, after meta-ultramafites. Laterite derived from Tertiary sedimentary rocks is well indurated at the top for about 2–5 metres (6.6–16.4 ft). Downwards, the profile grades into soft laterite with remnants of gritstone and culminates into a zone of variegated clay.

Chemical composition

It has been inferred from chemical analysis of Angadipuram laterites that they are a derivative of charnockite. The laterite which occurs at an average elevation of about 60 metres (200 ft) in the Angadipuram area have an admixture of pyroxene granulite, charnockite and migmatite. The results of the chemical analysis of samples of these laterites indicate the following composition.

- SiO_2 - 32%, Al_2O_3 – 29.38%, Fe_2O_3 –17.38%, TiO_2 – 2.05%, Na_2O – 0.95%, KO – 0.27%, CaO – 0.3% and MgO – 0.2%

Spatial variations have been recorded in the chemical composition of laterites in Kerala.

ECONOMIC USES

The economic importance of laterites comes from the mining of metals, particularly nickel and aluminium. Bauxite is an aluminium-rich laterite variety, which is commercially in demand. Bauxite patches are found with laterites overlying for a thickness varying from about 1–50 metres (3.3–164.0 ft). Worldwide resources inventory indicates that laterites are a major source of nickel and account for 70% of resources. But nickel production from this source is limited to about 40% of the total world production. Pepper one of the commercially important spices of Kerala is grown in red laterites as it provides well drained conditions with good water holding capacity. It is also rich in humus and essential plant nutrients.

Access

Angadipuram has a flourishing tourist industry because of its famous temples and is located about 16 kilometres (9.9 mi) from Malappuram on the road to Palghat. It is an important railway station on the Shoranur – Nilambur railway line. The main road from Palghat (Palakkad) to Calicut (Kozhikode) passes through Angadipuram viaPerinthalmanna. Perinthalmanna is the taluk headquarters located 1.5 kilometres (0.93 mi) from Angadipuram. The nearest airport is Calicut (Kozikode), 50 kilometres (31 mi) away.

SOIL MUSEUM

Soil properties and horizon development vary from place to place depending on climate, organisms, topographic position, parent material and time. Different forest ecosystem and other land covers make strong imprints on the soil beneath them and the information on these changes facilitates improved land management decisions that maintain soil productivity and therefore preserve forest sustainability and long-term ecosystem health.

Currently there are 20 Soil monoliths in the museum with depicts the variation in morphological properties of soil beneath different forest ecosystems in the Kerala part of Western Ghats

1. Soil profile from the evergreen forests
2. Located at an elevation of 600 m in the Sholayar range, Vazhachal Forest Division is protected from the intense actions of weathering agents by the thick vegetation. Surface layer is enriched with organic residues and the soil column extends only up to 56 cm above the parent rock.
3. Soil profile from Shola forests
4. Located at an elevation of 1920m in Munnar Forest Division is extra ordinarily black in colour throughout the depth due to the unique vegetation and cool climatic condition. This soil has a good reserve of organic carbon (5%) at 65cm depth
5. Soil profile from grasslands
6. Located at an elevation of 2120m in Eravikulam National Park has 5.7% organic carbon at 0-20cm depth. Even though the dense root mat is grasslands enriches the surface layer with organic carbon, back of canopy cover intensify the weathering process at deeper layers giving reddish colour

7. Soil profile from dry deciduous forests
8. Located at an elevation of 960m in Chinnar clearly reflect the rain shadow condition of the locality with pH of 7.3 in the surface layer and without any signs of laterisation in the deeper layers
9. Soil profile from scrub forests
10. Located at an elevation of 800m in Chinnar is in the early stage of its development with plenty of weathering rock pieces and poor reserve of organic carbon .
11. Soil profile from degraded forests
12. Located at an elevation of 60m in Pattikkad range. Thrissur Forest Division is unique with its relatively low organic carbon throughout the depth and hard laterite beneath

STATE SOIL MUSEUM

In tune with the growing concerns in the natural resource management front, Kerala government entrusted the establishment of a State Soil Museum to the Department of Soil Survey & Soil Conservation. The Department of Soil Survey and Soil Conservation is the premier institution dealing exclusively on soils of the State. Soil survey activities have helped to identify soils that occur extensively and that have special significance. The Soil Museum is an attempt to bring all these soils and their live models under one roof. Soil Museum and Soil Information Centre is established in the first floor of the Central Soil Analytical Laboratory at Parottukonam, Thiruvananthapuram. This museum is therefore the first of its kind in the world with the largest collection of soil monoliths and display materials. Potential user groups of soil museum would include farmers, soil scientists, research scholars, planners, administrators, students, extension personnel, non- governmental organizations and general public.

Objectives:

- Reference centre on the Soils of Kerala
- Display of soil monoliths of the Bench Mark Soils of the Major land Resource Areas of all districts
- Display of rocks and minerals of the State
- Reference base for planning and undertaking soil series based agricultural research activities in the State
- Mini theatre for providing information on soils and environmental issues at State and National Level through video shows
- Soil Information Centre displaying soil maps, watershed maps and related maps of all districts
- Display of watershed model, the basic unit of all developmental programs, and models of different soil and water conservation structures

BENCH MARK SOILS OF KERALA-CONCEPTS

MAJOR LAND RESOURCE AREAS

Early farmers realized that the different soils and climates they encountered required them to grow certain types of crops in order to survive economically. 'Major Land Resources Areas of Kerala' is the brainchild of Sri. Hari Eswaran, Scientist, USDA conceived during the First International Forum on Soil Taxonomy and Sustainable Land Management held in February-March 1993. The basic concept was built up on by the department personnel who identified 27 resource areas in Kerala based on physiography, geology, climate, soils and crops as parameters.

BENCH MARK SOILS

Eighty two Bench Mark soils have been identified in the 27 different Major Land Resource Areas of Kerala for which soil monoliths were collected. A Bench Mark Soil is one that is widely extensive, holds a key position in the soil classification system and is of special significance to farming, engineering or other uses.

The basic objective of Bench Mark Soil is to focus attention on extensively occurring soils and their agronomic concepts for wider acceptability of interpretations and for extrapolation of research data. The Bench Mark Soils are selected from among the existing and established soils that represent typical range of soil conditions within the state and having extensive coverage. It is the representative of the most extensive soils in major land resource area or agro ecological zone of the state and/or holding a key position in Soil Taxonomy. The selected soil should have large amount of research data for making transfer of agro technology feasible. Studies of Bench Mark Soils helps in soil correlation, standardization of legends, prediction of soil behaviour, agro technology transfer and planning further research in allied disciplines.

SOIL MONOLITHS

A Soil Monolith is a sample of a soil profile with undisturbed structure, including several or all the basic genetic horizons. The soil monoliths in the soil museum are arranged district-wise and based on Major land Resource Areas (MLRA's). Detailed information on soil series, their classification (Soil Taxonomy, 1994), landscape, crops, climate, typical profile characteristics fertility status etc are displayed.

SOIL INFORMATION CENTRE

Soil Information Centre is intended as a documentation and reference centre on the soils of Kerala. The center provides soil information in the form of hard copy, digital files of published maps and reports. The access and query facility on the soils of Kerala provides a detailed description on soil types, their limitations, ameliorative meaures etc. The success stories in the soil conservation front are also available for reference in the library.

ROCKS AND MINERALS

The common rocks and minerals seen in Kerala are displayed in the soil museum

WATERSHED MODEL

1. The word 'watershed' is derived from German 'Wasserscheide': wasser (water) & scheide (divide / parting)
2. Watershed is defined as a geo-hydrological unit draining to a common point by a system of drains. It is an area of land and water bounded by a drainage divide within which the surface run-off collects and flows out through a single outlet into a larger river or lake.
3. Each watershed is divided from the adjacent watersheds by a drainage divide usually the high elevation perimeter of that area (ridge line).
4. Size of the watershed depends upon the area covered by the selected drainage line, whereby smaller streams result in small area watersheds and larger streams give rise to larger area of the watersheds.
5. Watersheds are seen in different shapes. Circular, square, rectangle, oval, triangle etc are the common shapes identified.

Presently, the watershed is the basic unit for implementation and development of all agricultural schemes. A watershed model is displayed to highlight the concepts of watershed delineation and integrated management.

MINI THEATRE

The mini theatre shows videos on soil survey and soil conservation activities in the State. The state of the planet is documentarised stressing the need for conservation activities. The soil perspective is highlighted in all videos. Videos on environmental concerns are also shown.

SOIL AND WATER CONSERVATION MODELS

Soil and water conservation practices are highlighted through models. Detailed descriptions are also displayed.

DISTRICT-WISE SOIL AND WATERSHED MAPS

Display and documentation materials on Geology of Kerala, geological time periods, physiography, soils of kerala, watersheds of all 14 districts are put up. The departmental activites are also shown. The output of field soil survey activities is inventory reports and maps. The reconnissance soil survey maps of all districts in 1:50,000 scale is displayed along with prioritized district-wise microwatershed atlas. Each soil map gives the soil associations encountered in each district along with their physiography. The watershed atlas is prioritized, delineated and codified at micro level for use of local self government institutions.

WESTERN GHATS

The Western Ghats or Sahyadri are a mountain range that runs almost parallel to the western coast of the Indian peninsula, located entirely in India. It is a UNESCO World Heritage Site and is one of the eight "hottest hotspots" of biological diversity in the world. It is sometimes called the Great Escarpment of India.

The range runs north to south along the western edge of the Deccan Plateau, and separates the plateau from a narrow coastal plain, called Konkan, along the Arabian Sea. A total of thirty nine properties including national parks, wildlife sanctuaries and reserve forests were designated as world heritage sites - twenty in Kerala, ten in Karnataka, five in Tamil Nadu and four in Maharashtra.

The range starts near the border of Gujarat and Maharashtra, south of the Tapti river, and runs approximately 1,600 km (990 mi) through the states of Maharashtra, Goa, Karnataka, Kerala and Tamil Nadu ending at Kanyakumari, at the southern tip of India. These hills cover 160,000 km^2 (62,000 sq mi) and form the catchment area for complex riverine drainage systems that drain almost 40% of India. The Western Ghats block southwest monsoon winds from reaching the Deccan Plateau. The average elevation is around 1,200 m (3,900 ft).

The area is one of the world's ten "Hottest biodiversity hotspots" and has over 7,402 species of flowering plants,1814 species of non-flowering plants, 139 mammal species, 508 bird species, 179 amphibian species, 6000 insects species and 290 freshwater fish species; it is likely that many undiscovered species live in the Western Ghats. At least 325 globally threatened species occur in the Western Ghats.

GEOLOGY

The Western Ghats are the mountainous faulted and eroded edge of the Deccan Plateau. Geologic evidence indicates that they were formed during the break-up of the supercontinent of Gondwana some 150 million years ago. Geophysical evidence indicates that the west coast of India came into being somewhere around 100 to 80 mya after it broke away from Madagascar. After the break-up, the western coast of India would have appeared as an abrupt cliff some 1,000 m (3,300 ft) in elevation.

Basalt is the predominant rock found in the hills reaching a thickness of 3 km (2 mi). Other rock types found are charnockites, granite gneiss, khondalites, leptynites, metamorphic gneisses with detached occurrences of crystalline limestone, iron ore, dolerites and anorthosites. Residual laterite and bauxite ores are also found in the southern hills.

MOUNTAIN RANGES

The Western Ghats extend from the Satpura Range in the north, go south past Maharashtra, Goa, through Karnataka and into Kerala and Tamil Nadu. Major gaps in the range are the Goa Gap, between the Maharashtra and Karnataka sections, and the Palghat Gap on the Tamil Nadu and Kerala border between the Nilgiri Hills and the Anaimalai Hills.

The mountains intercept the rain-bearing westerly monsoon winds, and are consequently an area of high rainfall, particularly on their western side. The dense forests also contribute to the precipitation of the area by acting as a substrate for condensation of moist rising orographic winds from the sea, and releasing much of the moisture back into the air viatranspiration, allowing it to later condense and fall again as rain.

The northern portion of the narrow coastal plain between the Western Ghats and the Arabian Sea is known as the Konkan Coast or simply Konkan, the central portion is called Kanara and the southern portion is called Malabar region or theMalabar Coast. The foothill region east of the Ghats in Maharashtra is known as Desh, while the eastern foothills of the central Karnataka state is known as Malenadu. The largest city within the mountains is the city of Pune (Poona), in the Desh region on the eastern edge of the range. The Biligirirangan Hills lie at the confluence of the Western and Eastern Ghats.

Sahyadhris

The major hill range starting from the north is the *Sahyadhri* (*the benevolent mountains*) range. This range is home to many hill stations, including Matheran, Lonavala-Khandala, Mahabaleshwar, Panchgani, Amboli Ghat, Kudremukh and Kodagu. The range is known as *Sahyadri* in Maharashtra and Karnataka and as *Sahya Parvatam* in Kerala.

Nilgiris

The Nilgiri mountains are in Northwestern Tamil Nadu and are home to the town of Ooty. The Bili giri rangana Betta southeast of Mysore in Karnataka, meet the Shevaroys(Servarayan range) and Tirumala range farther east, linking the Western Ghats to the Eastern Ghats.

Anaimalai Hills

South of the Palghat Gap are the Anaimalai Hills, located in western Tamil Nadu and Kerala. There are smaller ranges further south, including the Cardamom Hills, thenAryankavu pass, Aralvaimozhi pass near Kanyakumari.

In the southern part of the range is Anamudi peak 2,695 metres (8,842 ft) in Kerala the highest peak in Western Ghats. Chembra Peak 2,100 metres (6,890 ft), Banasura Peak2,073 metres (6,801 ft), Vellarimala 2,200 metres (7,218 ft) and Agasthya mala 1,868 metres (6,129 ft) are also in Kerala. Doddabetta in the Nilgiri Hills is 2,637 metres (8,652 ft). Mullayanagiri is the highest peak in Karnataka 1,950 metres (6,398 ft). The Western Ghats in Kerala and Tamil Nadu is home to many tea and coffee plantations.

Fig. Shola Grasslands and forests in the Kudremukh National Park, Western Ghats, Karnataka.

Peaks

Following is a list of some of the highest peaks of the Western Ghats:

Rank	Name	Elevation (m)	Location
01.	Anaimudi	2695	Eravikulam National Park, Kerala
02.	Meesapulimala	2640	Eravikulam National Park, Kerala
03.	Doddabetta	2637	Nilgiris, Tamil Nadu
04.	Kolaribetta	2629	Mukurthi National Park, Tamil Nadu
05.	Mukurthi	2554	Mukurthi National Park, Tamil Nadu
06.	Vandaravu Peak	2553	Palani Hills Wildlife Sanctuary and National Park, Tamil Nadu
07.	Kattumala	2552	Eravikulam National Park, Kerala
08.	Anginda peak	2383	Silent Valley National Park, Kerala
09.	Vavul mala	2339	Vellarimala, Kerala
10.	Kodaikanal	2133	Kodaikanal, Tamil Nadu
11.	Chembra Peak	2100	Wayanad, Kerala
12.	Elivai Mala	2088	Palakkad, Kerala
13.	Banasura Peak	2073	Wayanad, Kerala
14.	Kottamala	2019	Periyar National Park, Kerala
15.	Mullayanagiri	1930	Chikmagalur, Karnataka
16.	Devarmala	1923	Achenkovil, Kerala
17.	Baba Budangiri	1895	Chikmagalur, Karnataka
18.	Kudremukh	1894	Chikmagalur, Karnataka
19.	Agasthyamala	1868	Thiruvananthapuram, Kerala
20.	Biligiriranga Hills	1800	Chamarajanagar, Karnataka
21.	Velliangiri Mountains	1778	Coimbatore, Tamil Nadu
22.	Tadiandamol	1748	Kodagu, Karnataka
23.	Kumara Parvata	1712	Dakshina Kannada, Karnataka

24.	Pushpagiri	1712	Pushpagiri Wildlife Sanctuary, Karnataka
25.	Kalsubai	1648	Ahmednagar, Maharashtra
26.	Brahmagiri	1608	Kodagu, Karnataka
27.	Salher	1567	Nashik, Maharashtra
28.	Madikeri	1525	Kodagu, Karnataka
29.	Dhodap	1472	Nashik, Maharashtra
30.	Himavad Gopala-swamy Betta	1450	Chamarajanagar, Karnataka
31.	Taramati	1431	Ahmednagar, Maharashtra
32.	Torna Fort	1405	Pune, Maharashtra
33.	Purandar fort	1387	Pune, Maharashtra
34.	Raigad fort	1346	Raigad, Maharashtra
35.	Kodachadri	1343	Shimoga, Karnataka.
36.	Paital Mala	1343	Kudiyanmala, Kerala
37.	Kote Betta	1620	Kodagu, Karnataka

LAKES AND RESERVOIRS

View from Varandha Pass showing the numerous waterfalls

The Western Ghats have several manmade lakes and reservoirs. The well known lakes are the Ooty (2500 m altitude, 34.0 ha) in Nilgiris, and the Kodaikanal (2285 m, 26 ha) and the Berijam in the Palani Hills. The Pookode lake of Wayanad in Kerala at Lakkadi is a beautiful scenic one with boating and garden arrangements. Most of the bigger lakes are situated in the state of Tamil Nadu. Two smaller lakes, the Devikulam (6.0 ha) and the Letchmi Elephant (2.0 ha) are in the Munnarrange.

The majority of streams draining the Western Ghats that join the rivers Krishna and Kaveri carry water during monsoon months only and have been dammed for hydroelectric and irrigation purposes. The major reservoirs are: Lonavala and Walwahn in Maharashtra; V.V. Sagar, K.R. Sagar, Bhadra reservoir at lakkavalli, linganamakki reservoir in the Malenadu area of Karnataka; Mettur Dam, Upper Bhavani, Mukurthi, Parson's Valley, Porthumund, Avalanche, Emerald, Pykara, Sandynulla, Karaiyar, Servalar, Kodaiyar, Manimuthar Dam and Glenmorgan in Tamil Nadu; and Kundallay and Maddupatty in the High Range of Kerala. Of these the Lonavla, Walwahn, Upper Bhavani, Mukurthi,

Parson's Valley, Porthumund, Avalanche, Emerald, Pykara, Sandynulla, Glenmorgan, Kundally and Madupatty are important for their commercial and sport fisheries for rainbow trout (introduced), mahseer (native) and common carp (introduced).

RIVERS

The Western Ghats form one of the four watersheds of India, feeding the perennial rivers of India. Important rivers include the Godavari, Tungabhadra, Krishna and Kaveri. These rivers flow to the east and drain out into the Bay of Bengal. The west flowing rivers, that drain into the Arabian Sea and the Laccadive Sea, are fast-moving, owing to the short distance travelled and steeper gradient. Important rivers include the Periyar, Bharathappuzha, Netravati, Sharavathi, Mandovi andZuari. Many of these rivers feed the backwaters of Kerala and Maharashtra. Rivers that flow eastwards of the Ghats drain into the Bay of Bengal. These are comparatively slower moving and eventually merge into larger rivers such as the Kaveri and Krishna. The larger tributaries include the Tunga River]], Bhadra river, Bhima River, Malaprabha River, Ghataprabha River, Hemavathi river, Kabini River. In addition there are several smaller rivers such as the Chittar River, Manimuthar River,Kallayi River, Kundali River and the Pachaiyar River.

Fig. Nilgiris a part of Western Ghats From Masinangudi

Fast running rivers and steep slopes have provided sites for many large hydro-electric projects. There are about 50 major dams along the length of the Western Ghats with the earliest project up in 1900 near Khopoli in Maharashtra. Most notable of these projects are the Koyna Hydroelectric Project in Maharashtra, the Parambikulam Dam in Kerala, and theLinganmakki Dam in Karnataka. The reservoir behind the Koyna Dam, the Shivajisagar Lake, has a length of 50 km(31 mi) and depth of 80 m (262 ft). It is the largest hydroelectric project in Maharashtra, generating 1,920 MW of electric power. Another major hydro electric project is Idukki dam in Kerala. This dam is one of the biggest in

Asia and generates around 70% of power for Kerala state. Mullaperiyar dam near Thekkady is one of the oldest in the world and a major tourist attractions in Kerala. Water from this dam is drawn to the vast coastal plain of Tamil Nadu, forming a delta and making it rich in vegetation.

Fig. The Jog Falls in Karnataka, one of the most spectacular waterfalls in India

During the monsoon season, numerous streams fed by incessant rain drain off the mountain sides leading to numerous and often spectacular waterfalls. Among the most well known is the Jog Falls, Kunchikal Falls, Dudhsagar Falls, Sivasamudram Falls, and Unchalli Falls. The Jog Falls is the highest natural plunge waterfall in South Asia and is listed among the 1001 natural wonders of the world. Talakaveri wildlife sanctuary is a critical watershed and the source of the river Kaveri. This region has dense evergreen and semi-evergreen vegetation, with shola-grassland in areas of higher elevation. The steep terrain of the area has resulted in scenic waterfalls along its many mountain streams. Sharavathi and Someshvara Wildlife sanctuaries in Shimoga district are the source of the Tungabhadra River system. The Netravathi river has also its origin at Western Ghats of India flowing westwards to join Arabian sea at Mangalore.

CLIMATE

Climate in the Western Ghats varies with altitudinal gradation and distance from the equator. The climate is humid and tropical in the lower reaches tempered by the proximity to the sea. Elevations of 1,500 m (4,921 ft) and above in the north and 2,000 m (6,562 ft) and above in the south have a more temperate climate. Average annual temperature here are around 15 °C (60 °F). In some parts frost is common, and temperatures touch the freezing point during the winter months. Mean temperature range from 20 °C (68 °F) in the south to 24 °C (75 °F) in the north. It has also been observed that the coldest periods in the South Western Ghats coincide with the wettest.

During the monsoon season between June and September, the unbroken Western Ghats chain acts as a barrier to the moisture laden clouds. The heavy, eastward-moving rain-bearing clouds are forced to rise and in the process deposit most of their rain on the windward side. Rainfall in this region averages 3,000–4,000 mm (120–160 in) with localised extremes touching 9,000 mm (350 in). The eastern region of the Western Ghats which lie in the rain shadow, receive far less rainfall averaging about 1,000 mm (40 in) bringing the average rainfall figure to 2,500 mm (150 in).

Data from rainfall figures reveal that there is no relationship between the total amount of rain received and the spread of the area. Some areas to the north in Maharashtra while receiving heavier rainfall are followed by long dry spells, while regions closer to the equator receiving less annual rainfall, have rain spells lasting almost the entire year.

ECOREGIONS

The Western Ghats are home to four tropical and subtropical moist broadleaf forest ecoregions – the North Western Ghats moist deciduous forests, North Western Ghats montane rain forests, South Western Ghats moist deciduous forests, andSouth Western Ghats montane rain forests.

The northern portion of the range is generally drier than the southern portion, and at lower elevations makes up the North Western Ghats moist deciduous forests ecoregion, with mostly deciduous forests made up predominantly of teak. Above 1,000 meters elevation are the cooler and wetter North Western Ghats montane rain forests, whose evergreen forests are characterised by trees of family *Lauraceae*.

The evergreen Wayanad forests of Kerala mark the transition zone between the northern and southern ecologic regions of the Western Ghats. The southern ecologic regions are generally wetter and more species-rich. At lower elevations are theSouth Western Ghats moist deciduous forests, with *Cullenia* the characteristic tree genus, accompanied by teak,dipterocarps, and other trees. The moist forests transition to the drier South Deccan Plateau dry deciduous forests, which lie in its rain shadow to the east.

Above 1,000 meters are the South Western Ghats montane rain forests, also cooler and wetter than the surrounding lowland forests, and dominated by evergreen trees, although some montane grasslands and stunted forests can be found at the highest elevations. The South Western Ghats montane rain forests are the most species-rich ecologic region in peninsular India; eighty percent of the flowering plant species of the entire Western Ghats range are found in this ecologic region.

BIODIVERSITY PROTECTION

Historically the Western Ghats were well-covered in dense forests that provided wild foods and natural habitats for nativetribal people. Its inaccessibility made it difficult for people from the plains to cultivate the land and build settlements. After the arrival of the British in the area, large swathes of territory were cleared for agricultural plantations and timber. The forest in the Western Ghats has been severely fragmented due to human activities, especially clear felling for tea, coffee, and teak plantations during 1860 to 1950. Species that are rare, endemic and habitat specialists are more adversely affected and tend to be lost faster than other species. Complex and species rich habitats like the tropical rainforest are much more adversely affected than other habitats.

Fig. A view of Ponmudi Hills in Thiruvananthapuram, Kerala

Fig. Western Ghats near Coimbatore, Tamil Nadu

The area is ecologically sensitive to development and was declared an ecological hotspot in 1988 through the efforts of ecologist Norman Myers. Though this area covers barely five percent of India's land, 27% of all species of higher plants in India (4,000 of 15,000 species) are found here. Almost 1,800 of these are endemic to the region. The range is home to at least 84 amphibian species, 16 bird species, seven mammals, and 1,600 flowering plants which are not found elsewhere in the world.

The Government of India established many protected areas including 2 biosphere reserves, 13 National parks to restrict human access, several wildlife sanctuaries to protect specific endangered species and many Reserve Forests, which are all managed by the forest departments of their respective state to preserve some of the ecoregions still undeveloped. Many National Parks were initially Wildlife Sanctuaries.

The Nilgiri Biosphere Reserve comprising 5500 km^2 of the evergreen forests of Nagarahole, deciduous forests of Bandipur National Park and Nugu in Karnataka and adjoining regions ofWayanad, Mudumalai National Park and Mukurthi National Park in the states of Kerala and Tamil Nadu forms the largest contiguous protected area in the Western Ghats. The Western Ghats is home to numerous serene hill stations likeMunnar, Ponmudi and Waynad. The Silent Valley National Park in Kerala is among the last tracts of virgin tropical evergreen forest in India.

Regarding the Western Ghats, in November 2009, Minister of Environment and Forests, Jairam Ramesh said,

"The Western Ghats has to be made an "ecologically sensitive zone". It is as important as the ecological system of the Himalayas for protection of the environment and climate of the country. The Central government will not give sanction for mining and hydroelectric projects proposed by the State Governments of Maharashtra, Karnataka and Goa that will destroy the Western Ghats eco-system."

In a letter dated 20 June 2009, Mr. Ramesh said,

"The (proposed) 200-MW Gundia hydel project of Karnataka Power Corporation in Hassan district would drown almost 1,900 acres (7.7 km^2) of thick forest in the already endangered Western Ghats along with all its fauna. This is something that both Karnataka and our country can ill-afford." "Power generation should not happen at the cost of ecological security."

The Expert Appraisal Committee appointed by Union Government also said that the project should not be taken up.

In August 2011, the Western Ghats Ecology Expert Panel (WGEEP) designated the entire Western Ghats as an Ecologically Sensitive Area (ESA) and, assigned three levels of Ecological Sensitivity to its different regions.

FAUNA

The Western Ghats are home to thousands of animal species including at least 325 globally threatened species. Many are endemicspecies, especially in the amphibian, reptilian and fish classes. Thirty two threatened species of mammals live in the Western Ghats. Of the 16 endemic mammals, 13 are threatened.

Mammals

There are at least 139 mammal species. A critically endangered mammal of the Western Ghats is the nocturnal Malabar large-spotted civet. The arboreal lion-tailed macaque is endangered. Only 2500 of this species are remaining. The largest population of lion tailed macaque is in Silent Valley National Park. Kudremukh National Park also protects a viable population.

These hill ranges serve as important wildlife corridors, allowing seasonal migration of endangered Asian elephants. The Nilgiri Bio-sphere is home to the largest population of Asian elephants and forms an important Project Elephant and Project Tiger reserve.

Brahmagiri and Pushpagiri wildlife sanctuaries are important elephant habitats. Karnataka's Ghat areas hold over six thousand elephants (as of 2004) and ten percent of India's critically endangered tiger population. The largest population of India's tigers outside the Sundarbans is in the forests where the boundaries of Karnataka, Tamil Nadu and Kerala meet.

The largest numbers and herds of vulnerable gaur are found here with the Bandipur National Park and Nagarhole together holding over five thousand gaur. To the west the forests of Kodagu hold sizeable populations of the endangered Nilgiri langur. Bhadra Wildlife Sanctuary and project tiger reserve in Lakkavalli of Chikmagalur has large populations of Indian muntjac. Many Asian elephant, gaur, sambar, vulnerable sloth bears, leopard, tiger and wild boars are found in the forests of Karnataka.

Bannerghatta National Park and Annekal reserve forest is an important elephant corridor connecting the forests of Tamil Nadu with those of Karnataka. Dandeli and Anshi national parks in Uttara Kannada district are home to leopards and significant populations of the great Indian hornbill. Bhimgad in Belgaum district is a proposed wildlife sanctuary and is home to the endemic critically endangered Wroughton's freetailed bat.

The Krishnapur caves close by are one of only three places in the country where the little-known Theobald's tomb bat is found. Large lesser false vampire bats are found in the Talevadi caves.

Reptiles

The snake family Uropeltidae of the reptile class is almost entirely restricted to this region.

Amphibians

The amphibians of the Western Ghats are diverse and unique, with more than 80% of the 179 amphibian species being endemic to the region. Most of the endemic species have their distribution in the rainforests of these mountains. The endangered purple frog was discovered in 2003 to be a living fossil. This species of frog is most closely related to species found in the Seychelles. Four new species of frogs belonging to the genera *Rhacophorus*, *Polypedates*, *Philautus* and *Bufo* were described from the Western Ghats in 2005. The region is also home to many caecilian species.

Fish

As of 2004, 288 freshwater fish species are listed for the Western Ghats, including 35 also known from brackish or marine water. Several new species have been described from the region since then (e.g., *Dario urops*, *Horabagrus melanosoma*, *Schistura kodaguensis* and *S. sharavathiensis*), meaning that the figure is higher today.

There are 118 endemic species, including 12 genera entirely restricted to the Western Ghats (*Betadevario*, *Dayella*, *Horabagrus*, *Horalabiosa*, *Hypselobarbus*,*Indoreonectes*, *Lepidopygopsis*, *Longischistura*, *Mesonoemacheilus*, *Parapsilorhynchus*, *Rohtee* and *Travancoria*).

There is a higher fish richness in the southern part of the Western Ghats than in the northern, and the highest is in the Chalakudy River, which alone holds 98 species. Other rivers with high species numbers include the Periyar, Bharatapuzha, Pamba andChaliyar, as well as upstream tributaries of the Kaveri, Pambar, Bhavani and Krishna rivers.

The most species rich families are theCyprinids (72 species), hillstream loaches (34 species; including stone loaches, now regarded a separate family), Bagrid catfishes (19 species) and Sisorid catfishes (12 species). The region is home to several brilliantly coloured ornamental fishes likeDenison's (or red line torpedo) barb, several species of *Dawkinsia* barbs, zebra loach, *Horabagrus* catfish, dwarf pufferfish anddwarf Malabar pufferfish. The rivers are also home to *Osteobrama bakeri*, and larger species such as the Malabar snakehead andMalabar mahseer.

According to the IUCN, 97 freshwater fish species from the Western Ghats were considered threatened in 2011, including 12 critically endangered, 54 endangered and 31 vulnerable. All but one (*Tor khudree*) of these are endemic to the Western Ghats. An additional 26 species from the region are considered data deficient (their status is unclear at present). The primary threats are fromhabitat loss, but also from overexploitation and introduced species.

Birds

There are at least 508 bird species. Most of Karnataka's five hundred species of birds are from the Western Ghats region.

Bhadra Wildlife Sanctuary is located at the northern end of the Malabar ranges and the southern tip of the Sahyadri ranges and bird species from both ranges can be seen here.

There are at least 16 species of birds endemic to the Western Ghats including the endangered rufous-breasted laughingthrush, the vulnerable Nilgiri wood-pigeon, white-bellied shortwing and broad-tailed grassbird, the near threatened grey-breasted laughingthrush, black-and-rufous flycatcher, Nilgiri flycatcher, and Nilgiri pipit, and the least concern Malabar (blue-winged) parakeet, Malabar grey hornbill, white-bellied treepie, grey-headed bulbul, rufous babbler, Wynaad laughingthrush, white-bellied blue-flycatcher and the crimson-backed sunbird.

Insects

There are roughly 6,000 insect species from Kerala alone. Of 334 Western Ghats butterfly species, 316 species have been reported from the Nilgiri Biosphere Reserve.

The Western Ghats is home to 174 species of odonates (107 dragonflies and 67 damselflies), including 69 endemics. Most of the endemic odonate are closely associated with rivers and streams, while the non-endemics typically are generalists.

Molluscs

Seasonal rainfall patterns of the Western Ghats necessitate a period of dormancy for its land snails, resulting in their high abundance and diversity including at least 258 species of gastropods from 57 genera and 24 families.

A total of 77 species of freshwater molluscs (52 gastropods and 25 bivalves) have been recorded from the Western Ghats, but the actual number is likely higher.

This include 28 endemics. Among the threatened freshwater molluscs are the mussels *Pseudomulleria dalyi*, which is a Gondwanan relict, and the snail *Cremnoconchus*, which is restricted to the spray zone of waterfalls. According to the IUCN, 4 species of freshwater molluscs are considered endangered and 3 are vulnerable. An additional 19 species are considered data deficients (their status is unclear at present).

FLORA

Of the 7,402 species of flowering plants occur in the Western Ghats, 5,588 species are native or indigenous and 376 are exotics naturalized and 1,438 species are cultivated or planted as ornamentals. Among the indigenous species, 2,253 species are endemic to India and of them, 1,273 species are exclusively confined to the Western Ghats.

Apart from 593 confirmed subspecies and varieties; 66 species, 5 subspecies and 14 varieties of doubtful occurrence are also reported and therefore amounting 8,080 taxa of flowering plants.

Western Ghats Ecology Expert Panel

The Western Ghats Ecology Expert Panel, headed by ecologist Madhav Gadgil, was a committee appointed by the Union Ministry of Environment and Forests to assess the biodiversity and environmental issues of the Western Ghats spread across six states-Kerala, Tamil Nadu, Karnataka, Goa, Maharashtra and Gujarat. The panel which was set up on 14 March 2010, submitted its report to the government on 31 August 2011. Gadgil Committee and its successor Kasturirangan Committee recommended suggestions to protect the Western Ghats. However, both of them ran into controversy and were not implemented.

Gadgil report was criticised for being too environment-friendly and not in tune with the ground realities. The Kasturirangan Committee tried to balance development and environment, but was labelled as being anti-environmental. Both reports had major methodological flaws that prevented objective assessment of ecologically sensitive areas (ESAs).

10

Trees

Kerala is a green state having variety of trees. The trees of Kerala are grown in forests, villages and in cityscapes. Trees are very useful in various ways where it provides food, medicine, timber, firewood etc. Trees also have many medicinal values. Some of the main trees found in our state include the following

THENGU (COCONUT TREE)

Coconut tree is the state tree of Kerala. In Kerala, coconut tree is called Kalpavriksha and is a tall and branchless tree. The fruit of the tree is big and oval shaped. Inside the fruit lies the thick brown fibre which surrounds the hard shell. The sweet and the pleasant edible material and the milk stay within this shell. It is used in the preparation of almost all kinds of dishes in Kerala. Oil extracted from 'Copra' obtained by drying the edible part is used for cooking, for making shampoos and can be applied to the hair for its enrichment. After the oil is extracted the copra is used as a food for fowls and cattle. Oil is used in the manufacture of soaps and candles. Coir is the solid skin of the nut and is used for making mats and strong ropes which are durable in salt water. Tender coconut water is a natural healthy drink and is good for dysentery, vomiting etc. Toddy obtained from coconut tree is a beverage. The leaves are used for thatching and the trunk is used for roof beam and bridges. The hard cases of the nuts are used for manufacturing cups and vessels.

COCONUT TREE FACTS

Coconut tree is a plant that belongs to the family Arecaceae. There are over 150 species of coconuts that can be found in 80 different countries throughout the world. Coconut tree grows only in the tropical climate. This plant live on the sandy soil, requires a lot of sunlight and regular rainfalls. Coconut tree does not tolerate low temperatures and low percent of humidity. Cultivated plants are prone to insect attacks which can decrease production of fruit worth of hundreds of million dollars. Coconut is important part of human diet because it contains valuable vitamins and minerals. Other than that, coconuts are used in the production of various wooden items, in the construction industry, in the manufacture of beauty products and as a fuel.

Interesting Coconut tree Facts:

Coconut trees can be dwarf and tall. Dwarf coconut trees can reach 20 to 60 feet in height, while tall coconut trees grow to the height of 98 feet.

Coconut tree has 13 to 20 inches long pinnate leaves. Leaflets have lanceolate shape. They can reach 24 to 35 inches in length.

Coconut tree is attached to the ground via strong fibrous root system.

Coconut tree develops male and female flowers. They mature at different time to avoid self-pollination. Two types of flowers can be distinguished by size: female flowers are larger.

Fruit of a coconut tree is botanically known as drupe. Fruit becomes fully ripe after one year. Under optimal weather conditions, coconut tree can produce 75 fruits per year, but that happens extremely rare. More often, coconut tree produces 30 fruits each year.

Ripe coconut has 3.2 pounds of weight. White, edible flesh of coconut can be consumed raw or dried. Dry version of coconut flesh is called "copra". Production of one ton of copra requires 6000 coconuts.

Each year, 61 million tons of coconuts are produced and distributed throughout the world. In certain parts of the world, people use trained monkeys to collect coconuts instead of them.

Coconut is very popular and often consumed fruit. Other than the flesh, coconut water, milk and oil can be used in the preparation of various healthy dishes. Coconut oil has wide application in cosmetic industry. Coconut oil is part of various lotions because it smells nice and increases the moisture of the skin. Sap extracted from the closed flower is used for the production of alcoholic beverage called "coconut vodka".

Wooden parts of coconuts are used in the manufacture of furniture, decorative objects, drums, containers and even canoes.

All parts of the coconuts can be exploited, even their hard shells. Husk and shell are discarded as waste after extraction of the fruit. They are used as fuel and source of charcoal. Coconut oil is used as a substitute for diesel fuel.

Coconuts can survive up to 100 years in the wild.

GROWING (HOW TO GROW A COCONUT TREE)

If you plan to grow coconuts, better have patience, seven years of patience before you can expect to see any coconuts. The coconut palm, Cocos nucifera, is considered a "three generation tree," supporting a farmer, his children, and his grandchildren. Some trees, which can grow to a height of 60 to 100 feet, even survive all three generations. Cocos nucifera, nucifera meaning nut-bearing, has only one species that includes both the tall and the dwarf coconut tree, but many varieties exist within the species.

Though the coconut is commonly considered a nut, botanically it is classified as a drupe and is the largest of all fruit seeds. The coconut consists of the thin, strong outer layer or skin called the epicarp, the thick fibrous layer called mesocarp, and the dark brown hard shell called the endocarp that encases the coconut flesh. Just beneath the endocarp is the testa, the thin deep brown layer that clings to the white coconut meat. The coconut palm is a striking tree with a tall slender trunk that keeps its same diameter from the base to the top. Beautiful, lacy fronds, about 25 to 35 of them, form an umbrella-like structure at the tree's zenith. The tree grows taller by forming new fronds that sprout from the top of the tree as the lower fronds die off.

Coconuts grow from the center of the fronds, close to the trunk. Unique to the coconut palm, each tree blooms thirteen times a year and produces all stages of growth at the same time, from tiny new green nuts to fully ripened brown nuts that are ready to fall from the tree.

Coconuts are persnickity about where they live. They cannot survive cold climates, and do poorly in temperate zones. Coconuts require the hot, humid weather of the tropical regions that stretch 25 degrees north to 25 degrees south of the equator all around the globe. There the sun shines steadily with plenty of rainfall to nurture the slow-growing coconut palm. To begin the growing process, purchase a coconut with its husk completely intact. Just like sprouting any seeds and legumes, the coconut must be soaked in water, only longer, two or three days.

Next, prepare a pot that is large enough and deep enough for the coconut by putting big pieces of gravel or stones in the bottom to allow for good drainage. Add about two inches of sandy soil, then set the coconut on the soil with the pointed or bud end up. Add more soil until it covers about half the coconut. Then set the pot in a warm place such as a sunny window, near a warm oven, or on a radiator. The next step requires patience and diligence. Pour warm water over the coconut husk every day, making sure it does not dry out. The sprouting process is very slow, sometimes taking six months or longer. Until the sprout appears, the coconut is receiving its nourishment from the white meat inside. The coconut water within provides the nut with all of its moisture requirements.

COCONUT PALM

The coconut palm is a palm tree in the family Arecaceae (palm family). It is a large palm, growing to 30 m tall. It has leaves that are 4–6 m long. The term *coconut* refers to the fruit of the coconut palm. The coconut tree is a monocot.

There are many coconut palms on the coasts of India and Bangladesh. People of this area use coconut milk in cooking. Women use coconut oil as oil for their hair. The coconut's shell is relatively hard, but can be broken. Because its shell is hard, it can be used as an ingredient to make craftworks.

Fig. Coconut from Ivory Coast

Coconut milk is also used in many drinks. Coconut oil is often in food and soaps. People in Sri Lanka use coconut flowers for wedding celebrations. In the Maldives it is the National tree.

A coconut is a large nut. Coconuts grow in tropical countries. The flesh of a coconut is white and can be eaten raw or used in cooking. It is used in many of the foods we eat for flavour. It is native to tropical areas.

Seed dispersal

Coconuts floats on water and can float to another island and germinate there. It has often been noted that coconuts can travel 110 days, or 3000 miles, by sea and still be able to germinate.

ATHI (CLUSTER FIG TREE)

This large deciduous tree grows in evergreen forests, moist places, near streams etc. It is often cultivated in villages and its edible fruit is known for its nutrient richness. These are trees with aerial roots. Barks are used for rinderpest disease seen in cattle. Athi is one of the four trees in Nalpamara, a prime mix of ayurveda includes athi, ithi, arrayal and peraal. Root used in diarrhea and diabetes. Fruit stomachic and carminative used in hemoptysis. Latex is used for cracks, piles and diarrhea.

Ficus racemosa

Ficus racemosa is a species of plant in the Moraceae family. Popularly known as the Cluster Fig Tree,Indian Fig Tree or Goolar (Gular) Fig, this is native to Australia, Malaysia, South-East Asia and the Indian Subcontinent. It is unusual in that its figs grow on or close to the tree trunk, termed cauliflory. In India the tree and its fruit are called *gular* in the north and *atti* in the south. The fruits are a favourite staple of the common Indian macaque. It serves as a food plant for the caterpillars of the butterfly the Two-brand Crow (*Euploea sylvester*) of northern Australia.

In Hinduism

In the Atharva Veda, this fig tree (Sanskrit: *u□umbara* or *udumbara*) is given prominence as a means for acquiring prosperity and vanquishing foes.

For instance, regarding an amulet of the *udumbara* tree, a hymn (AV xix,31) extols:

The Lord of amulets art thou, most mighty: in thee wealth's ruler hath engendered riches,

These gains are lodged in thee, and all great treasures. Amulet,

conquer thou: far from us banish malignity and indigence, and hunger.

Vigour art thou, in me do thou plant vigour: riches art thou, so do thou grant me riches.

Plenty art thou, so prosper me with plenty: House-holder, hear a householder's petition.

It has been described in the story of Raja Harischandra of the Ikshvaku dynasty, that the crown was a branch of this Udumbura tree, set in a circlet of gold. Additionally, the Throne (simhasana) was constructed out of this wood and the royal personage would ascend it on his knee, chanting to the gods to ascend it with him, which they did so, albeit unseen. The tree has been worshipped as Abode under which Guru Duttatreya Dutta,teaches that to teach first learn from others however small or Big.there is always something to be learnt from One and to learn new things one has to learn to Unlearn as per the time.The tree is seen planted in all the places associated with Lord Duttatreya who is seen as an icon Rishi a sage who represents all the three of the TRINITY of hinduism-Brahma, Vishnu and Shiva,Creator,maintainer,and destroyer needed for each one to learn by unlearning the obsolete.this is the plan of evolution in analogy.

In Buddhism

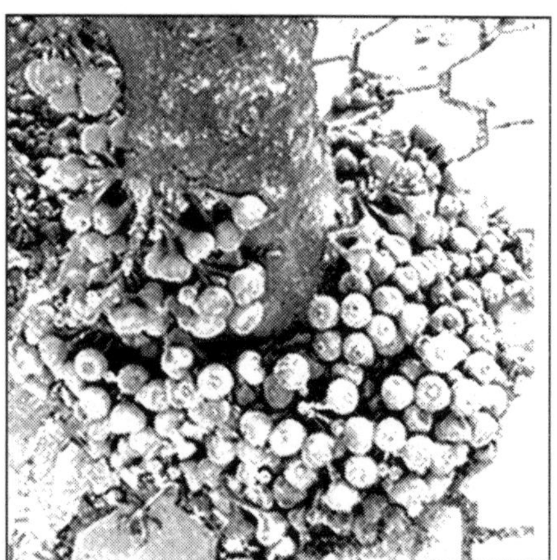

Fig. Clusters of gular figs on a tree trunk in India

Both the tree and the flower are referred to as the udumbara (Sanskrit, Pali; Devanagari: !A.M,0) in Buddhism. Udumbara can also refer to the blue lotus (*Nila udumbara*) flower. The udumbara flower appears, an importantMahayana Buddhist text. The Japanese word *udonge* was used by Dôgen Zenji to refer to the flower of the udumbara treeof the *Shôbôgenzô* ("Treasury of the Eye of the True Dharma"). Dôgen places the context of the udonge flower in the Flower Sermon given by Gautama Buddha on Vulture Peak. *Udonge* is also used to refer to the eggs of the lacewing insect. The eggs are laid in a pattern similar to a flower, and its shape is used for divination in Asian fortune telling.

Uses

Fig. Pickled and halved gular figs

Fig. Lion-tailed macaque feeding on this fig

In ancient times both Hindu and Buddhist ascetics on their way to Taxila, (Original name is Taksha Sila) travelling through vast areas ofIndian forests used to consume the fruit during their travels. One challenge to vegetarians were the many fig wasps that one finds when opening a gular One way to get rid of them was to break the figs into halves or quarters, discard most of the seeds and then place the figs into the midday sun for an hour. Gular fruit are almost never sold commercially because of this problem .

The Ovambo people call the fruit of the Cluster Fig *eenghwiyu* and use it to distill *Ombike*, their traditional liquor.

Health Uses

The bark of Audumbar/Oudumbar tree is said to have healing power. In countries like India, the bark is rubbed on a stone with water to make a paste and the paste is applied over the skin which is afflicted by boils or mosquito bites. Allow the paste to dry on the skin and reapply after a few hours. For people whose skin is especially sensitive to insect bites; this is a very simple home remedy.

MEDICINAL USE OF GULAR TREE OR CLUSTER FIG TREE

Gular (Ficus racemosa , syn. Ficus glomerata Roxb.) tree is available throughout India. It is also known as Cluster fig in English. Goolar tree leaves, fruits and bark are used to treat various diseases.

Gular tree is astringent, antidiabetic, antiasthma tic, anti-inflammatory, antioxidant, antiulcer, anti-pyretic and anti diarrheal in action. Its latex is used to treat piles and hemorrhoids. Bark of gular tree is used to treat infections, swelling and inflammation.

Mouth ulcers and other mouth infections

- Take bark (10 gm)of gular tree and cook in water(400 ml) for few minutes. Add alum(pinch) and filter. Use this water to rinse mouth frequently .

Swelling

- Take bark of gular tree, grind on stone with water to make paste. Apply at affected area.

Boils

- For curing boils take gular bark paste and apply at affected area.

Pimples and freckles

- Take inner side of gular bark and make paste of it. Apply on pimples, acne and freckles.

Burn marks on skin

- Take gular fruits and make paste. Mix paste with honey and apply regularly on burn marks. Use regularly to get normal skin tone.
- Also burning causes skin to tighten and stiffen. Apply paste of bark and leaves at such burnt areas.

Rakta Pitta , Nose Bleeding

- Take tender leaves(10-15), grind to extract juice and add misri to this juice. Take regularly.

Urticaria(Hives) or sheetpitta

- Take tender leaves and extract juice. Drink 15 ml juice for few days.

Bhasmak rog

- Dry gular fruits and make powder. Take this powder(1 tbsp) three times a day for few days.

Dysentery

- Drink 10- 15 ml juice of gular tender leaves.

Stomach ache

- Take dry fruits of gular, ajwain and sendha namak. Make powder and take two times a day with water.

Fistula, Piles Hemorrhoids

- Take milk like secretion or latex of gular tree after plucking leaves and soak in cotton. Apply on affected area.

Leucorrhea, Weakness , Spermatorrhea

- Drink 1 cup juice of gular tender leaves in morning.

Burning In Hands And Feet Due To Excess Pitt

- Drink tender leaves juice of gular.

ARAYAAL (PEEPAL)

It is also called as Arasu or Bodhi and it is of great medicinal values. Peepal is closely linked with Indian culture. Most of the temple premises have vast spreading peepal trees, which was once a platform for social gatherings of devotees. It is considered as a sacred tree and its leaves are used for religious purposes. It is believed that Lord Krishna passed away under this tree after which the Kaliyuga had begun. Lord Buddha attained enlightenment meditating under the Peepal tree. Peepal leaves have numerous benefits as a result of which it is used in ayurveda. It is especially useful for patients suffering from jaundice to control urine. For constipation problem, there can be no better remedy. Peepal leaves can help to get rid of mumps. Having very thick leaf strength, peepal contributes much to strengthen the ozone layer releasing large amount of oxygen at any time.

Ficus religiosa

Ficus religiosa or sacred fig is a species of fig native to Nepal, India, Bangladesh, Myanmar, Pakistan, Sri Lanka, south-west China andIndochina. It belongs to the *Moraceae*, the fig or mulberry family. It is also known as the bodhi tree, pippala tree, peepal tree orashwattha tree (in India and Nepal).

DESCRIPTION

Ficus religiosa is a large dry season-deciduous or semi-evergreen tree up to 30 metres (98 ft) tall and with a trunk diameter of up to 3 metres (9.8 ft). The leaves are cordate in shape with a distinctive extended drip tip; they are 10–17 cm long and 8–12 cm broad, with a 6–10 cm petiole. The fruits are small figs 1–1.5 cm in diameter, green ripening to purple. As an interesting fact, the leaves of this tree move continuously even when the air around is still and no wind is blowing. This phenomenon can be explained due to the long leaf stalk and the broad leaf structure. However, religious minded people in Hindu/ Buddhist religion attribute this movement of the leaves to the fact that "devas" or "gods" reside on these leaves and make it move continuously. This fact is also mentioned in the Bhagavad Gita as a verse "O Ashvatha, I honor you whose leaves are always moving..." Another unique quality of this tree is it releases oxygen day and night, make it even more significant and respected where it exists.

IN RELIGION

The *Ficus religiosa* tree is considered sacred by the followers of Hinduism, Jainism and Buddhism. In the Bhagavad Gita, Krishna says, "I am the Peepal tree among the trees, Narada among the sages, Chitraaratha among the Gandharvas, And sage Kapila among the Siddhas."

Buddhism

Gautama Buddha attained enlightenment (*bodhi*) while meditating underneath a *Ficus religiosa*. The site is in present-day Bodh Gaya in Bihar, India. The original tree was destroyed, and has been replaced several times. A branch of the original tree was rooted inAnuradhapura, Sri Lanka in 288 BCE and is known as Jaya Sri Maha Bodhi; it is the oldest flowering plant (angiosperm) in the world.

In Theravada Buddhist Southeast Asia, the tree's massive trunk is often the site of Buddhist or animist shrines. Not all *Ficus religiosa* can be called a *Bodhi Tree*. A Bodhi Tree must be able to trace its parent to another Bodhi Tree and the line goes on until the first Bodhi Tree under which Gautama is said to have gained enlightenment.

Hinduism

Sadhus (Hindu ascetics) still meditate beneath sacred fig trees, and Hindus do pradakshina (circumambulation, or meditative pacing) around the sacred fig tree as a mark of worship. Usually seven pradakshinas are done around the tree in the morning time chanting "*vriksha rajaya namah*", meaning "salutation to the king of trees."

It claimed that the 27 stars (constellations) constituting 12 houses (*rasis*) and 9 planets are specifically represented precisely by 27 trees—one for each star. The Bodhi Tree is said to represent Pushya(Western star name ã, ä and è Cancri in the Cancer constellation).

Plaksa is a possible Sanskrit term for *Ficus religiosa*. However, according to Macdonell and Keith (1912), it denotes the wavy-leaved fig tree (*Ficus infectoria*) instead. In Hindu texts, the Plaksa tree is associated with the source of the Sarasvati River. The *Skanda Purana*states that the Sarasvati originates from the water pot of Brahma flows from Plaksa on the Himalayas. According to Vamana Purana32.1-4, the Sarasvati was rising from the Plaksa tree (Pipal tree). *Plaksa Pra-sravana* denotes the place where the Sarasvati appears. In the Rigveda Sutras, Plaksa Pra-sravana refers to the source of the Sarasvati.

CLASSIFICATION

Other names: Bo tree, Bodhi tree, Sacred tree, Beepul tree, Pipers, Pimpal, Jari, Arani,Ashvattha, Ragi, Bodhidruma, Shuchidruma, Pipalla, Ashvattha and the Buddha tree are the other names used for the Banyan tree.

Kingdom :	Plantae
Division	Magnoliophyta
Class:	Magnoliopsida
Order :	Rosales
Family :	Moraceae
Genus :	Ficus
Species :	F. religiosa
Scientific Name :	Ficus religiosa
Found In :	**Ranthambore Wildlife Sanctuary**

Description: Peepal is a large, fast growing deciduous tree. It has a heart shaped leaves. It is a medium size tree and has a large crown with the wonderful wide spreading branches. It shed its leaves in the month of March and April. The fruits of the Peepal are hidden with the figs. The figs are ripen in the month of May. The figs which contain the flowers grow in pairs just below the leaves and look like the berries. Its bark is light gray and peels in patches. Its fruit is purple in colour. It is one of the longest living trees.

Other Species: Artocarpus heterophyllus Lam, Artocarpus incissus L., Artocarpus nobilis Thw. Are some of the other species of the Peepal tree.

Location: Peepal tree is grown throughout India. It is mainly grown in State of Haryana, Bihar, Kerala and Madhya Pradesh. It is also found in the Ranthambore National Park in India.

Cultivation: Peepal tree is easily propagated through the seeds or through the cuttings. It can grow in any type of soil. Young peepal needs proper nourishment. It requires full sunlight and proper watering.

Medicinal uses: This tree of life has also got the medicinal value. The juice of its leaves extracted by holding them near the fire can be used as the ear drop. Its power bark has been used to heal the wounds for years. The bark of the tree is useful in inflammations and glandular swelling of the neck. Its root bark is useful for stomatitis, clean ulcers, and promotes granulations. Its roots are also good for gout. The roots are even chewed to prevent gum diseases. Its fruit is laxative which promotes digestion and checks vomiting. Its ripe fruits are good for the foul taste, thirst and heart diseases. The powered fruit is taken for Asthma. Its seeds have proved useful in urinary troubles. The leaves are used to treat constipation.

Other uses: People in India collect the Peepal leaves, clean them, dry them and than paint them with the gold acrylic in order to preserve them for years. From the bark of the Peepal tree reddish dye is extracted. Its leaves are used to feed the camels and the elephants. When the leaves are dried they are used for the decoration purpose.

Cultural importance: Peepal tree has the great importance in India especially among the Buddhist who regard Peepal tree as the personification of Buddha. Lord Buddha attained enlightenment mediating under the Peepal tree. It is regarded as the sacred tree and the people uses its leaves for the religious purposes. According to the Buddha – 'He who worships the Peepal tree will receive the same reward as if he worshiped me in person'. The Peepal tree has its own symbolic meaning of Enlightenment and peace. People tie threads of white, red and yellow silk around it to pray for progeny and rewarding parenthood. Hindus in India holds the great spiritual regard for the Peepal Tree, they regard it as the tree beneath which Vishnu was born.

ASOKAM (ASOKA TREE)

The flowers are flame coloured, fragrant and are considered that of kamadeva, God of romance in Hindu mythology. The skin of the bark, leaves, fruits and flowers of Asoka are used for medicinal purpose. It is used for the treatment of uterine bleeding, dysmenorrheal-painful menstruation, bleeding piles, and many other diseases. This herb is rarely used externally and when used, the paste of its bark-skin is good for fighting pain associated with swelling. It can be used internally by women for treating various gynecological disorders. Asoka blooms in the summer and produces fruits in October. The fruit is bitter and astringent in taste, pungent in post digestive effect and has a cold potency. It helps to reduce vata and kapha dosas.

POLYALTHIA LONGIFOLIA

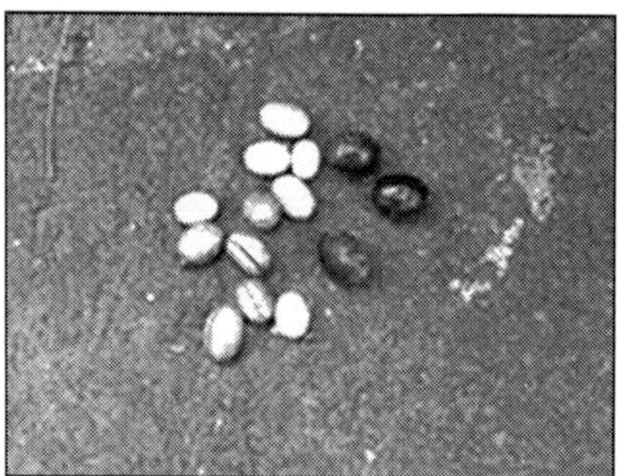

Fig. Seeds and fruit of *Polyalthia longifolia*

Polyalthia longifolia (False Ashoka) is a lofty evergreen tree, native to India, commonly planted due to its effectiveness in alleviating noise pollution. It exhibits symmetrical pyramidal growth with willowy weeping pendulous branches and long narrow lanceolate leaves with undulate margins. The tree is known to grow over 30 ft in height.

Polyalthea is derived from a combination of Greek words meaning 'many cures' with reference to the medicinal properties of the tree while Longifolia, in Latin, refers to the length of its leaves. Polyalthia longifolia is sometimes incorrectly identified as the Ashoka tree (Saraca indica) because of the close resemblance of both trees.

Fig. Polyalthia longifolia

One might mistake it as a tree with effectively no branches, but in fact a *Polyalthia* allowed to grow naturally (without trimming the branches out for decorative reasons) grows into a normal large tree with plenty of shade.

COMMON NAMES

Polyalthia longifolia's common names include False Ashoka, the Buddha Tree, Indian mast tree, and Indian Fir tree.

Distribution

Found natively in India and Sri Lanka. It is introduced in gardens in many tropical countries around the world. It is, for example, widely used in parts of Jakarta in Indonesia.

Leaves

Fresh leaves are a coppery brown color and are soft and delicate to touch, as the leaves grow older the color becomes a light green and finally a dark green. The leaves are shaped like a lance and have wavy edges. The leaves are larval food plant of the Tailed Jay and the Kite Swallowtail butterflies.

FLOWERING

In spring the tree is covered with delicate star-like pale green flowers. The flowers last for a short period, usually two to three weeks, are not conspicuous due to their color.

Fruit is born in clusters of 10-20, initially green but turning purple or black when ripe. These are loved by birds such as the Asian koel *Eudynamys scolopaceus* and bats including flying foxes.

Fig. Close up flowers inHyderabad, India.

Uses

The leaves are used for ornamental decoration during festivals. The tree is a main attraction in gardens throughout India. The tree can be cut into various shapes and maintained in required sizes. In past, the flexible, straight and light-weight trunks were used in the making of masts for sailing ships. That is why the tree is also known as the Mast Tree. Today, the tree is mostly used for manufacturing small articles such as pencils, boxes, matchsticks, etc.

Methanolic extracts of *Polyalthia longifolia* have yielded 20 known and two new organic compounds, some of which show cytotoxic

ASHOKA TREE BENEFITS

Ashoka tree is one of the sacred and legendary trees of India. Asoka is one of the important medicinal trees. The tree is revered by Buddhists, Hindus and Jains. The tree's medicinal importance has been inscribed in various Ayurvedic scriptures. All parts of the tree are used in medicine.

Ayurveda has made the best use of Ashoka tree. The Ayurvedic properties of Ashoka are soothing, cooling, dry and pungent. The meaning of the word Ashoka means the a person without any worries.

Health Benefits of Ashoka

Some medicinal benefits of Ashoka are as follows:

Menstruation – The bark of the tree is used to cure excessive loss of blood during menstruation in presence of leucorrhea, uterine fibroids and other reasons. The bark is taken as a decoction and used as a substitute for a dried fungus, Ergot used to cure the uterine hemorrhgaes.

Dysentery – For treatment of dysentery, decoction is used. Fluid extract of Ashoka flowers is used to treat hemorrhagic dysentery. This fluid is prepared by grinding its flowers with water. The extract is taken in the doses of 15 to 60 drops. Diabetes – Dried flowers of Ashoka are used for diabetes treatment. Piles – Bark of the tree is used for cure of internal piles. Decoction prepared from bark is used to cure piles. Insect Bite – The bark of the tree is used to treat scorpion bite. Complexion Enhancement – The bark of the tree is used for enhancing complexion. Blood Purification – Ashoka is used to prepare Ayurvedic tonics to purify blood. Worms Infestation – The leaves and bark are used to get rid of worms in stomach.

Ashokarishta is an Ayurvedic preparation from Ashoka tree to cure various diseases in women. It reduces excessive bleeding, leucorrhea and headache.

VEPPU (NEEM)

Neem is popularly known as divine tree. Ability to purify air makes it an inevitable part of courtyard trees. The neem tree is drought resistance and is found in areas with sub-arid to sub-humid conditions where annual rainfall between 400 and 1200mm. It grows in regions where annual rainfall is below 400mm and in such cases it depends upon ground water levels. Products made from neem have several medicinal properties such as antifungal, antidiabetic, antibacterial and many others. It is a major ingredient in ayurvedic medicine and is prescribed for skin disease. It is a source of environment-friendly biopesticides.

Azadirachta indica

Azadirachta indica, also known as Neem, Nimtree, and Indian Lilac is a tree in the mahogany family Meliaceae. It is one of two species in the genus *Azadirachta*, and is native to India and the Indian subcontinent including Nepal, Pakistan, Bangladesh and Sri Lanka. It is typically grown in tropical and semi-tropical regions. Neem trees now also grow in islands located in the southern part of Iran. Its fruits and seeds are the source of neem oil.

Description

Neem is a fast-growing tree that can reach a height of 15–20 metres (49–66 ft), rarely to 35–40 metres (115–131 ft). It is evergreen, but in severe drought it may shed most or nearly all of its leaves. The branches are wide and spreading. The fairly dense crown is roundish and may reach a diameter of 15–20 metres (49–66 ft) in old, free-standing specimens. The neem tree is very similar in appearance to its relative, the Chinaberry (*Melia azedarach*).

The opposite, pinnate leaves are 20–40 centimetres (7.9–15.7 in) long, with 20 to 31 medium to dark green leaflets about 3–8 centimetres (1.2–3.1 in) long. The terminal leaflet is often missing. The petioles are short.

The (white and fragrant) flowers are arranged in more-or-less drooping axillary panicles which are up to 25 centimetres (9.8 in) long. Theinflorescences, which branch up to the third degree, bear from 150 to 250 flowers. An individual flower is 5–6 millimetres (0.20–0.24 in) long and 8–11 millimetres (0.31–0.43

in) wide. Protandrous, bisexual flowers and male flowers exist on the same individual tree. The fruit is a smooth (glabrous) olive-like drupe which varies in shape from elongate oval to nearly roundish, and when ripe is 1.4–2.8 centimetres (0.55–1.10 in) by 1.0–1.5 centimetres (0.39–0.59 in). The fruit skin (exocarp) is thin and the bitter-sweet pulp (mesocarp) is yellowish-white and very fibrous. The mesocarp is 0.3–0.5 centimetres (0.12–0.20 in) thick. The white, hard inner shell (endocarp) of the fruit encloses one, rarely two or three, elongated seeds (kernels) having a brown seed coat.

Etymology

Neem is a Hindi noun derived from Sanskrit Nimba.

Vernacular names

Names for this plant in various languages include;

- English - Margosa, Neem Tree
- Hindi - Neem
- Manipuri - Neem
- Marathi - Nimbay
- Tamil - Veppai, Sengumaru
- Malayalam - Ariyaveppu
- Telugu - Vepa
- Kannada - Turakabevu
- Bengali - Neem
- Punjabi - Nimm
- Urdu - Neem
- Assamese - Neem
- Gujarati - Dhanujhada, Limba
- Sanskrit - Pakvakrita, nimbaka
- Sinhala - Kohomba
- Yoruba - Dongoyaro

Ecology

The neem tree is a kind of tree noted for its drought resistance. Normally it thrives in areas with sub-arid to sub-humid conditions, with an annual rainfall 400–1,200 millimetres (16–47 in). It can grow in regions with an annual rainfall below 400 mm, but in such cases it depends largely on ground water levels. Neem can grow in many different types ofsoil, but it thrives best on well drained deep and sandy soils. It is a typical tropical to subtropical tree and exists at annual mean temperatures between 21–32 °C (70–90 °F). It can tolerate high to very high temperatures and does not tolerate temperature below 4 °C (39 °F). Neem is one of a very few shade-giving trees that thrive in drought-prone areas e.g. the dry coastal, southern districts of India and Pakistan. The trees are not at all delicate about water quality and thrive on the merest trickle of water, whatever the quality. In India and tropical countries where the Indian diaspora has reached, it is very common to see neem trees used for shade lining streets, around temples, schools & other such public buildings or in most people's back yards. In very dry areas the trees are planted on large tracts of land.

Weed status

Neem is considered a weed in many areas, including some parts of the Middle East, most of Sub-Saharan Africa including West Africa and Indian Ocean states, and some parts of Australia. Ecologically, it survives well in similar environments to its own, but its weed potential has not been fully assessed.

In April 2015, *A. indica* was declared a class B and C weed in the Northern Territory, Australia, meaning its growth and spread must be controlled and plants or propagules are not allowed to be brought into the NT. It is illegal to buy, sell, or transport the plants or seeds. Its declaration as a weed came in response to its invasion of waterways in the "Top End" of the territory.

After being introduced into Australia, possibly in the 1940s, *A. indica* was originally planted in the Northern Territory to provide shade for cattle. Trial plantations were established between the 1960s and 1980s in Darwin, Queensland, and Western Australia, but the Australian neem industry did not prove viable. The tree has now spread into the savanna, particularly around waterways, and naturalised populations exist in several areas.

Uses

Fig. Neem tree

Fig. Neem tree in the Philippines

Fig. Neem tree leaves in india

Neem leaves are dried in India and placed in cupboards to prevent insects eating the clothes and also while storing rice in tins. Neem leaves are dried and burnt in the tropical regions to keep away mosquitoes. These flowers are also used in many Indian festivals like Ugadi.

As a vegetable

The tender shoots and flowers of the neem tree are eaten as a vegetable in India. A souplike dish called *Veppampoo charu* (Tamil) (translated as "neem flower rasam") made of the flower of neem is prepared in Tamil Nadu. In West Bengal, young neem leaves are fried in oil with tiny pieces of eggplant (brinjal). The dish is called *nim begun* and is the first item during a Bengali meal that acts as an appetizer. It is eaten with rice.

Neem is used in parts of mainland Southeast Asia, particularly in Cambodia aka sdov—, Laos (where it is called *kadao*), Thailand(where it is known as *sadao* or *sdao*), Myanmar (where it is known as *tamar*) and Vietnam (where it is known as *s□u ðâu* and is used to cook the salad *g□i s□u ðâu*). Even lightly cooked, the flavour is quite bitter and the food is not enjoyed by all inhabitants of these nations, though it is believed to be good for one's health. Neem gum is a rich source of protein. In Myanmar, young neem leaves and flower buds are boiled with tamarind fruit to soften its bitterness and eaten as a vegetable. Pickled neem leaves are also eaten with tomato and fish paste sauce in Myanmar.

Traditional medicinal use

Products made from neem trees have been used in India for over two millennia for their medicinal properties. Neem products are believed by Siddha and Ayurvedic practitioners to be anthelmintic, antifungal, antidiabetic, antibacterial, antiviral, contraceptive andsedative. It is considered a major component in siddha medicine and Ayurvedic and Unani medicine and is particularly prescribed forskin diseases. Neem oil is also used for healthy hair, to improve liver function, detoxify the blood, and balance blood sugar levels. Neem leaves have also been used to treat skin diseases like eczema, psoriasis, etc. However, insufficient research has been done to assess the purported benefits of neem. In adults, short-term use of neem is safe, while long-term use may harm the kidneys or liver; in small children, neem oil is toxic and can lead to death. Neem may also cause miscarriages, infertility, and low blood sugar.

Safety issues

Neem oil can cause some forms of toxic encephalopathy and ophthalmopathy if consumed in large quantities.

Pest and disease control

Neem is a key ingredient in non-pesticidal management (NPM), providing a natural alternative to synthetic pesticides. Neem seeds are ground into a powder that is soaked overnight in water and sprayed onto the crop. To be effective, it is necessary to apply repeatedly, at least every ten days. Neem does not directly kill insects on the crop. It acts as an anti-feedant, repellent, and egg-laying deterrent, protecting the crop from damage. The insects starve and die within a few days. Neem also suppresses the hatching of pest insects from their eggs. Neem cake is often sold as a fertilizer.

Neem oil has been shown to avert termite attack as an ecofriendly and economical agent. In the UK, plant protection products that contain azadirachtin, the active ingredient of neem oil, are illegal.

Neem oil for polymeric resins

Applications of neem oil in the preparation of polymeric resins have been documented in the recent reports. The synthesis of various alkyd resins from neem oil is reported using a monoglyceride (MG) route and their utilization for the preparation of PU coatings. The alkyds are prepared from reaction of conventional divalent acid materials like phthalic and maleic anhydrides with MG of neem oil. In other reports, different routes for preparation of polymeric resins from neem oil are also reported.

Other uses

The twigs are also used as tooth brush. One end is chewed to turn it into soft bristles to clean the teeth by brushing.

- Toiletries: Neem oil is used for preparing cosmetics such as soap, shampoo, balms and creams as well as toothpaste.
- Toothbrush: Traditionally, slender neem twigs (called datun;) are first chewed as a toothbrush and then split as a tongue cleaner. This practice has been in use in India, Africa, and the Middle East for centuries. Many of India's 80% rural population still start their day with the chewing stick, while in urban areas neem toothpaste is preferred. Neem twigs are still collected and sold in markets for this use, and in rural India one often sees youngsters in the streets chewing on neem twigs. It has been found to be equally effective as a toothbrush in reducing plaque and gingival inflammation.
- Tree: Besides its use in traditional Indian medicine, the neem tree is of great importance for its anti-desertification properties and possibly as a good carbon dioxide sink.
- Neem gum is used as a bulking agent and for the preparation of special purpose foods.
- Neem blossoms are used in Andhra Pradesh, Tamil Nadu and Karnataka to prepare Ugadi pachhadi. A mixture of neem flowers andjaggery (or unrefined brown sugar) is prepared and offered to friends and relatives, symbolic of sweet and bitter events in the upcoming new year, Ugadi. "*Bevina hoovina gojju*" (a type of curry prepared with neem blossoms) is common in Karnataka throughout the year. Dried blossoms are used when fresh blossoms are not available. In Tamil Nadu, a rasam (veppam poo rasam) made with neem blossoms is a culinary specialty.
- Cosmetics: Neem is perceived in India as a beauty aid. Powdered leaves are a major component of at least one widely used facial cream. Purified neem oil is also used in nail polish and other cosmetics.
- Bird repellent: Neem leaf boiled in water can be used as a very cost effective bird repellent, especially for sparrows.
- Lubricant: Neem oil is non drying and it resists degradation better than most vegetable oils. In rural India, it is commonly used to grease cart wheels.

- Fertilizer: Neem has demonstrated considerable potential as a fertilizer. Neem cake is widely used to fertilize cash crops, particularly sugarcane and vegetables. Ploughed into the soil, it protects plant roots from nematodes and white ants, probably as it contains the residual limonoids. In Karnataka, people grow the tree mainly for its green leaves and twigs, which they puddle into flooded rice fields before the rice seedlings are transplanted.
- Resin: An exudate can be tapped from the trunk by wounding the bark. This high protein material is not a substitute for polysaccharide gum, such as gum arabic. It may however, have a potential as a food additive, and it is widely used in South Asia as "Neem glue".
- Bark: Neem bark contains 14% tannin, an amount similar to that in conventional tannin yielding trees (such as Acacia decurrens). Moreover, it yields a strong, coarse fibre commonly woven into ropes in the villages of India.
- Honey: In parts of Asia neem honey commands premium prices, and people promote apiculture by planting neem trees.
- Soap: 80% of India's supply of neem oil is now used by neem oil soap manufacturers. Although much of it goes to small scale speciality soaps, often using cold-pressed oil, large scale producers also use it, mainly because it is cheap. Additionally it is antibacterial and antifungal, soothing and moisturising. It can be made with up to 40% neem oil. Well known brands include Margo. Generally, the crude oil is used to produce coarse laundry soaps.
- Against pox viruses: In Tamil Nadu, people who are affected with pox viruses are generally made to lie in bed made of neem leaves and branches. This prevents the spreading of pox virus to others and has been in practice since early centuries.

ASSOCIATION WITH HINDU FESTIVALS IN INDIA

Neem leaf or bark is considered an effective pitta pacifier due to its bitter taste. Hence, it is traditionally recommended during early summer in Ayurveda (that is, the month ofChaitra as per the Hindu Calendar which usually falls in the month of March – April).

In the Indian states of Telangana, Andhra Pradesh and Karnataka, Neem flowers are very popular for their use in 'Ugadi Pachhadi' (soup-like pickle), which is made on Ugadiday. In Telangana, Andhra Pradesh and Karnataka, a small amount of Neem and Jaggery (Bevu-Bella) is consumed on Ugadi day, the Telugu and Kannada new year, indicating that one should take both bitter and sweet things in life, joy and sorrow.

During Gudi Padva, which is the New Year in the state of Maharashtra, the ancient practice of drinking a small quantity of neem juice or paste on that day, before starting festivities, is found. As in many Hindu festivals and their association with some food to avoid negative side-effects of the season or change of seasons, neem juice is associated with Gudi Padva to remind people to use it during that particular month or season to pacify summer pitta.

In Tamil Nadu during the summer months of April to June, the Mariamman temple festival is a thousand year old tradition. The Neem leaves and flowers are the most important part of the Mariamman festival. The goddess Mariamman statue will be garlanded with Neem leaves and flowers. During most occasions of celebrations and weddings the people of Tamil Nadu adorn their surroundings with the Neem leaves and flowers as a form of decoration and also to ward off evil spirits and infections. In the eastern coastal state of Odisha the famous Jagannath temple deities are made up of Neem heart wood along with some other essential oils and powders.

CHEMICAL COMPOUNDS

Ayurveda was the first to bring the anthelmintic, antifungal, antibacterial, and antiviral constituents of the Neem tree to the attention of natural products chemists. The process of extracting neem oil involves extracting the water-insoluble components with ether, petrol ether,ethyl acetate and dilute alcohol. The provisional naming was *nimbin* (sulphur-free crystalline product with melting point at 205 °C,empirical composition $C_7H_{10}O_2$), *nimbinin* (with similar principle, melting at 192 °C), and *nimbidin* (cream-coloured containing amorphoussulphur, melting at 90–100 °C). Siddiqui identified *nimbidin* as the main active antibacterial ingredient, and the highest yielding bitter component in the neem oil. These compounds are stable and found in substantial quantities in the Neem. They also serve as natural insecticides.

Neem coated urea is being used an alternate to plain urea fertilizer in India. It reduces pollution, improves fertilizer's efficacy and soil health.

Genome and transcriptomes

Neem genome and transcriptomes from various organs have been sequenced, analyzed and published by Ganit Labs in Bangalore, India. ESTs were identify by generation of subtractive hybridization libraries of neem fruit, leaf, fruit mesocarp and fruit endocarp by CSIR-CIMAP Lucknow.

CULTURAL AND SOCIAL IMPACT

In Theravada Buddhism, the neem tree is said to have been used to achieve enlightenment (*bodhi*) by Tissa, the twentieth reincarnation of Lord Buddha. However, some sources claim that *Terminalia tomentosa* was the Bodhi tree used. In 1995, the European Patent Office (EPO) granted a patent on an anti-fungal product derived from neem to the United States Department of Agriculture and W. R. Grace and Company. The Indian government challenged the patent when it was granted, claiming that the process for which the patent had been granted had actually been in use in India for over 2,000 years. In 2000, the EPO ruled in India's favour but W. R. Grace appealed, claiming that prior art about the product had never been published in a scientific journal. On 8 March 2005, that appeal was lost and the EPO revoked the Neem patent.

NEEM

Neem (*Azadirachta indica*) is a tree in the mahogany family Meliaceae. It is native to India, Myanmar, Bangladesh, Sri Lanka,Malaysia and Pakistan. It grows in tropical and semi-tropical regions. It was once the state tree of Osmanistan. Neem is a fast-growing tree that can reach up to 15–20 m (about 50–65 feet) tall, and sometimes even to 35–40 m (115–131 feet). It is evergreen, but in serious drought it may lose most or nearly all of its leaves. The branches are spread far apart.

Uses

Products made from neem have been used in India for over two millennia for their medicinal properties: they are said to be antifungal, antidiabetic, antibacterial, antiviral, contraceptive and sedative. Neem products are also used in selectively controllingpests in plants. Neem is considered a large part ofAyurvedic medicine.

- All parts of neem are used for preparing many different medicines, especially for skin disease.
- Part of the Neem tree can be used as a spermicide .
- Neem oil is used for preparing cosmetics (soap and shampoo, as well as lotions and others), and is useful for skin care such asacne treatment. Neem oil has been used effectively as a mosquito repellent.
- Neem is useful for damaging over 500 types of insects, mites, ticks, and nematodes, by changing the way they grow and act. Neem does not normally kill pests right away, rather it slows their growth and drives them away. As neem products are cheap and not poisonous to animals and friendly insects, they are good for pest control

ELANJI (WEST INDIAN MEDLAR)

Elanji is a tropical tree that produces tiny flowers having sweet aroma during night. The leaves form a part of perfumes with intoxicating fragrance. The fruit is edible, has a thick outer skin which encloses sweet yellow pulp and a hard shell seed. Different parts of the tree have medicinal properties. In Kerala, people are very familiar to Elanjithara melam, very scintillating and highly appreciated percussion held at Elanjithara in the Vadakkumnatha temple compound during Thrissur Pooram.

Importance: Spanish cherry, West Indian Medlar or Bullet wood tree is an evergreen tree with sweet- scented flowers having ancient glamour. Garlands made of its flowers are ever in good demand due to its long lasting scent. Its bark is used as a gargle for odontopathy, ulitis and ulemorrhagia. Tender stems are used as tooth brushes. It is also useful in urethrorrhoea, cystorrhoea, diarrhoea and dysentery. Flowers are used for preparing a lotion for wounds and ulcers. Powder of dried flowers is a brain tonic and is useful as a snuff to relieve cephalgia. Unripe fruit is used as a masticatory and will help to fix loose teeth. Seeds are used for preparing suppositories in cases of constipation especially in children. The bark and seed coat are used for strengthening the gum and enter into the composition of various herbal tooth powders, under the name of "Vajradanti", where they may be used along with tannin-containing substances like catechu (Acacia catechu), pomegranate (Punica granatum) bark, etc. The bark is used as snuff for high fever accompanied by pains in various parts of the body. The flowers are considered expectorant and smoked in asthma. A lotion prepared from unripe fruits and flowers is used for smearing on sores and wounds. In Ayurveda, the important preparation of Mimusops is "Bakuladya Taila", applied on gum and teeth for strengthening them, whereas in Unani system, the bark is used for the diseases of genitourinary system of males.

Distribution: It is cultivated in North and Peninsular India and Andaman Islands. It is grown as an avenue tree in many parts of India.

Botany: Mimusops elengi Linn. belongs to the family Sapotaceae. It is an evergreen tree with dark grey fissured bark and densely spreading crown. Leaves are oblong, glabrous and leathery with wavy margins. Flowers are white, fragrant, axillary, solitary or fascicled. Fruits are ovoid or ellipsoid berries. Seeds are 1-2 per fruit, ovoid, compressed, greyish brown and shiny. Other important species belonging to the genus Mimusops are M. hexandra Roxb. and M. kauki Linn. syn. Manilkara kauki Dub.

Agrotechnology: Mimusops prefers moist soil rich in organic matter for good growth. The plant is propagated by seeds. Fruits are formed in October-November. Seeds are to be collected and dried. Seeds are to be soaked in water for 12 hours without much delay and sown on seedbeds. Viability of seeds is less. After germination they are to be transferred to polybags. Pits of size 45cm cube are to be taken and filled with 5kg dried cowdung and top soil. To these pits, about 4 months old seedlings from the polybags are to be transplanted with the onset of monsoon. Addition of 10kg FYM every year is beneficial. Any serious pests or diseases do not attack the plant. Flowering commences from fourth year onwards. Bark, flowers, fruit and seeds are the economic parts.

Properties and activity: -sitosterol and its glucoside, -spina-sterol, quercitol, taraxerol and lupeol and its acetate are present in the aerial parts as well as the roots and seeds. The aerial parts in addition gave quercetin, dihydroquercetin, myricetin, glycosides, hederagenin, ursolic acid, hentriacontane and -carotene. The bark contained an alkaloid consisting largely of a tiglate ester of a base with a mass spectrum identical to those of laburinine and iso-retronecanol and a saponin also which on hydrolysis gave -amyrin and brassic acid. Seed oil was comprised of capric, lauric, myristic, palmitic, stearic, arachidic, oleic and linoleic acids.

Saponins from seed are spermicidal and spasmolytic. The aerial part is diuretic. Extract of flower (1mg/kg body weight) showed positive diuretic action in dogs. Bark is tonic and febrifuge. Leaf is an antidote for snakebite. Pulp of ripe fruit is antidysenteric. Seed is purgative. Bark and pulp of ripe fruit is astringent.

Mimusops elengi

Fig. The ripe fruit has many traditional uses.

Mimusops elengi is a medium-sized evergreen tree found in tropical forests in South Asia, Southeast Asia and northern Australia. English common names include Spanish cherry, medlar, and bullet wood.

Its flower is the provincial flower of Yala Province, Thailand.

Tree description

Fig. Flowers in Hyderabad, India

Bullet wood is an evergreen tree reaching a height of about 16 m (52 ft). It flowers in April, and fruiting occurs in June. Leaves are glossy, dark green, oval-shaped, 5–14 cm (2.0–5.5 in) long, and 2.5–6 cm (0.98–2.36 in) wide. Flowers are cream, hairy, and scented. Bark is thick and appears dark brownish black or grayish black in colour, with striations and a few cracks on the surface.

The tree may reach up to a height of 9–18 m (30–59 ft) with about 1 m (3 ft 3 in) in circumference.

Ayurvedic uses

The bark, flowers, fruits, and seeds of *Bakula* are used in Ayurvedic medicine in which it is purported to be astringent, cooling,anthelmintic, tonic, and febrifuge. It is mainly used for dental ailments such as bleeding gums, pyorrhea, dental caries, and loose teeth.

Other uses

- The edible fruit is softly hairy becoming smooth, ovoid, bright red-orange when ripe.
- The wood is a luxurious wood that is extremely hard, strong and tough, and rich deep red in color. The heart wood is sharply defined from the sapwood. It works easily and takes a beautiful polish. Weight is 1008 kg per cubic meter.

KANIKKONNA (GOLDEN SHOWER TREE)

The tree is an indicator of rain, blooming before the onset of monsoon. Parts of the tree such as roots, bark and fruit pulp are often used in ayurvedic medicines. The bark of the tree is rich in tannins and is used in leather processing and fabric colouring.

THE GOLDEN SHOWER

Without doubt, this is a lovely flowering tree. Indian Laburnum or morecommonly known as, Amaltas is one of the most beautiful of our indigenous trees; it adds colour to our hills during the drier and hotter parts of summer. The beauty of the picture encompassing the extremely beautiful, bright yellow flowers of the tree is beyond words!

During the hot season when the long, drooping sprays of clcar, yellow flowers clothe the tree in a mantle of gold, it isindeed a glorious sight. Each spray is more than 30 cm. in length and bears long, slim stalks with numerous, large, deliciously fragrant flowersand rounded buds. Theseflowers are attractive to bees, butterflies and small birds, making it an extremely showy tree in bloom.

The awesome tree is equally beautiful! In Ayurvedic medicine, Golden Shower tree is known *aragvadha*, meaning 'disease killer'. Its fruit pulp is used as a mild laxative, used against fevers, arthritis, *vatavyadhi*(nervous system diseases), all kinds of *rakta-pitta* (bleeding, such as hematemesis or hemorrhages), as well as cardiac conditions and stomach problems such as acid reflux. The root is considered a very strong purgative and self-medication or any use without medical supervision is strongly advised against in Ayurvedic texts.

Amaltas is the state flower of Kerala. The flowers are of ritual importance in the*Vishu* festival of Kerala which is considered as the first day of Zodiac calender. The *Vishukkani* is inseparable from *Vishu*. According to the age-old belief, an auspicious *kani*(first sight) at dawn on the*Vishu* day is lucky for the entire year. As a result, the*Vishukkani* is prepared with a lot of care to make it the most positive sight so as to bring alive a wonderful, propitious and year ahead! The feast on the *Vishu* day also include food items made from Neem andMango.

There is an interesting fact related to the fruits of Amaltas. The fruits are dark-brown cylindrical in shape, 2 feet in length. The pod is hard enough and there in no way that the seeds will be dispersed after drying of the pods or so. So how, do you think, does the natural seed dispersal of Amaltas take place? Well Nature has it all! The seeds are enclosed in sweet pulp which has laxative properties. In forests, it is the main attraction for sloth bears and jackals. They eat the pod and seeds pass through their Gastro-Intestinal tract. Then the seeds are dropped out at different places by these animals. :-)

Amaltas is propagated from seed. Artificially, soaking in boiling water for 5 minutes and then 24 hour cold water soaking can help in germinating the seeds faster.

The tree is very useful in Apiculture. It is often utilised as shade tree or windbreak. It also gives excellent quality charcoal and tannin. The leaves provide useful green manure. Amaltas is widely planted in gardens and societies due its ornamental properties. It makes an excellent show when planted along the roads. The *Laburnum Road* in Mumbai has been exclusively lined with Amaltas trees and hence named so..!

KANJIRAM (NUXVOMICA)

Basically Kanjiram is a perennial tree. It is medium-sized with short, thick, trunk. It can be used in the form of tonic, stimulant and febrifuge. The useful parts of the plant are seed, bark, roots and leaves. The plant is also good to pacify vitiates kapha, vata, hypotension, arthritis and dementia. In large doses all part of the tree is toxic. Kanjiram is good for upset stomach, constipation, hangovers, heartburn, certain heart diseases, circulatory problems, depression, headaches, nervous conditions and respiratory diseases. Thus, kanjiram is a one end solution for almost all intestinal and nervous disorders.

ILAVU (SILK COTTON TREE)

It is a deciduous tree with a straight trunk and spreading branches. Flowers are red in colour and the seeds contain stable oils. It is used to cure vata and pitta disorders. It helps in relieving from burning sensation and it removes mark caused due to burning. It is also a remedy for dysentery, diarrhea, cough, cold and general body weakness.

KARINOCHI (FIVE LEAVED CHASTE TREE)

Karinochi is a large aromatic shrub whose leaves and roots have medicinal properties. The juice of the leaves have anti inflammatory, anti bacterial, anti fungal and analgesic properties. It is also used in the treatment of bruises, injuries and other skin infections. Leaves are used like tea for getting relief from cough and asthma. The root is used to cure the dyspepsia, colic, rheumatism, worms and leprosy. The flowers are used for curing diseases like diarrhea, cholera, fever and diseases of liver. The seeds are a cooling medicine for various skin diseases.

KARIVELAM (ACACIA)

It is a small to moderate sized thorny tree that grows up to 10 m in height and has golden flowers. Bark and gum are useful part of the tree.

EZHILAM PALA (DEVIL TREE)

Ezhilam pala, the name itself makes one to think of 'Yakshi', is a moderate to large sized evergreen tree which grows up to 30 m in height. The leaves come out in whorls of seven, elliptic-oblong, obtuse and petiolate, flowers small greenish white, found in umbellate cymes, fruits follicles about 50 cm long, contain papillose seeds with hairs on each ends. The blooms are associated with very strong aroma. The soft wood is used for manufacturing packing cases, match sticks, black boards, pencils etc.

KOOVALAM (WOOD APPLE)

Its mythological links to Lord Shiva makes 'Koovalam' a sacred plant to Hindus. The garland made of koovalam leaves is special to Lord Shiva and this tree is planted in the temple premises. Very much known for its medicinal properties, unripe or half ripe fruits are used as astringent, digestive etc. Its cooled aromatic pulp can be used to cure diabetes, inflammation, diarrhea, constipation etc. The fruits are edible and the bark of the tree is toxic.

Limonia acidissima

Limonia acidissima is the only species within the monotypic genus *Limonia*. It is native in the Indomalaya ecozone to Bangladesh,India, Pakistan, Sri Lanka, and in Indochinese ecoregion east to Java and the Malesia ecoregion. Vernacular names in English include:wood-apple, elephant-apple, monkey fruit, and curd fruit; and listed below are the variety of common names in the languages of itsnative habitat regions.

DESCRIPTION

Fig. Wood-apple tree in Trincomalee, Sri Lanka

Limonia acidissima is a large tree growing to 9 metres (30 ft) tall, with rough, spiny bark. The leaves are pinnate, with 5-7 leaflets, each leaflet 25–35 mm long and 10–20 mm broad, with a citrus-scent when crushed. The fruit is a berry 5–9 cm diameter, and may be sweet or sour. It has a very hard rind which can be difficult to crack open, and contains sticky brown pulp and small white seeds. The fruit looks similar in appearance to fruit of Bael (Aegle marmelos).

Uses

The rind of the fruit is so thick and hard it can be carved and used as a utensil such as a bowl or ashtray. The bark also produces anedible gum. The tree has hard wood which can be used for woodworking. Bael fruit pulp has a soap-like action that made it a household cleaner for hundreds of years. The sticky layer around the unripe seeds is household glue that also finds use in jewellery-making. The glue, mixed with lime, waterproofs wells and cements walls. The glue also protects oil paintings when added as a coat on the canvas.

Ground *Limonia* bark is also used as a cosmetic called thanakha in Myanmar, a practice that has spread to other parts of Southeast Asia. The fruit rind yields oil that is popular as a fragrance for hair; it also produces a dye used to colour silks and calico. It is a hedge plant favored for its rapid growth; especially when cuttings from a faster-growing individual are grafted to a hardily rooted plant, fruit, foliage and shade can quickly be obtained.

In Tamil Nadu leaves and fruit traditionally have been used for elephant food, while the branches were used as brooms for rough work in connection with animal care.

Culinary

The fruit is eaten plain, blended into an assortment of drinks and sweets, or well-preserved as jam. The scooped-out pulp from its fruits is eaten uncooked with or without sugar, or is combined with coconut milk and palm-sugar syrup and drunk as a beverage, or frozen as an ice cream. It is also used in *chutneys* and for making jam. A drink, Bael-panna made by blending the fruit with water and spices, is drunk during summers.

Indonesians beat the pulp of the ripe fruit with palm sugar and eat the mixture at breakfast. The sugared pulp is a foundation of sherbet in the subcontinent. Jam, pickle, marmalade, syrup, jelly, squash and toffee are some of the foods of this multipurpose fruit. Young bael leaves are a salad green in Thailand. Indians eat the pulp of the ripe fruit with sugar or jaggery. The ripe pulp is also used to make chutney. The raw pulp is varied with yoghurt and make into raita. The raw pulp is bitter in taste, while the ripe pulp has a smell and taste that is a mixture of sourness and sweet.

To those unfamiliar with this fruit, its unique and overpowering scent may make it almost unpalatable. Some have described it as smelling rotten or fermented to such an extent that some doubt it is even edible.

Nutrition

A hundred grams of fruit pulp contains 31 grams of carbohydrate and two grams of protein, equivalent to nearly 140 calories. The ripe fruit is rich in beta-carotene, a precursor of Vitamin A; it also contains significant quantities of the B vitamins thiamine and riboflavin, and small amounts of vitamin C.

KARINGALI (CUTCH)

It is a perennial tree which quenches severe thirst and has high medicinal value. It is a medium sized thorny deciduous tree with beautiful flowers. Karingali is good for diabetic patients. Its stem is used for brushing teeth to protect gums and teeth from decay. Drinking water boiled with dried karingali powder has brownish colour and it acts as a good appetizer. Karingali is an essential ingredient of arishtam and kashayam, good for curing the illness of cough, itching etc. It can be planted as an ornamental tree.

NENMENI VAKA (EAST INDIAN WALNUT)

Nenmeni vaka is a medium sized deciduous tree, makes a haven of pinkish white or yellowish white fragrant flowers. Bark, flowers and seed of the plant pacifies vitiated pitta, kapha, urinary retention etc. It is used to produce timber.

CHOONDAPPANA (WINE PALM)

Choondappana is one of the commonest wild palms on the Western Ghats. Its ornamental feature makes an attractive pot plant at its young age. The stem of the tree is smooth and similar to cylinders in shape. The fruits are reddish in colour. The wood is much stronger and durable than the other palm trees and is useful for domestic purposes. Toddy can be obtained by tapping. If you remove the fatal bud or the cabbage of the tree, it will kill the tree. Fibre that is made from the sheaths of the leaf stalks is very strong and can be used in making ropes, brushes etc.

Jubaea

Jubaea chilensis (Chilean wine palm or Chile cocopalm) is the sole extant species in the genus *Jubaea* in the palm familyArecaceae. It is native to southwestern South America, where it is endemic to a small area of central Chile, between 32°S and 35°S in southern Coquimbo, Valparaíso, Santiago, O'Higgins and northern Maule regions. It was long assumed that the extinct palm tree ofEaster Island belonged to this genus too, but it is distinct and now placed in its own genus, *Paschalococos*. It is a palm reaching heights of 25 metres (82 ft) with a trunk up to 1.3 metres (4.3 ft) in diameter at the base, often thicker higher up, and with smooth bark. The 3–5-metre (9.8–16.4 ft) leaves are pinnate. The largest individual specimen of indoor plant in the world is the*Jubaea chilensis* at Kew Gardens, England.

Etymology

The genus was named after Juba II, a Berber king and botanist.

Growth

The tree grows very slowly, as it is usual for palm trees. It takes several years until the Jubaea starts getting its weight and size. It may take more than 20 years for the plant to get the height of a medium tree.

Ecology

It needs mild winters, but will tolerate frosts down to about "15 °C (5 °F) as well as relatively cool summers, making it one of the hardiestof pinnate-leaved palms; this is because it grows up to 1,400 metres (4,600 ft) above sea level in its natural habitat. In the wild, the tree lives almost exclusively on the steep slopes of ravines. In the U.S. this palm grows best in the west from Seattle, Wa. south to San Diego, Ca. also east through Arizona, New Mexico, and west Texas, this palm suffers and can die in areas with extreme heat combined with high humidity.

Economic uses

Fig. Fruits and nuts

The common name refers to the past use of the sap from the trunk of this palm to produce a fermented beverage. The sap is also boiled down into a syrup and sold locally as *miel de palma*.

The tree also produces small round fruits that are about 2–3 centimetres (0.79–1.18 in) in diameter. The fruit has a very hard outer shell and whitish meat on the inside, like a miniature coconut. The fresh nuts are commonly sold in the areas where the palms grow during their fruiting season.

Conservation

The species is partially protected within Chile, although pressures of human overpopulation and expansion of grazing area have reduced the population of the Chilean Wine Palm in recent centuries. Unlike most other palm wines, collecting the sap requires cutting down the tree; this harvesting also has reduced the population of *Jubaea*.

PALM WINE

Palm wine is an alcoholic beverage created from the sap of various species of palm tree such as the palmyra, date palms, and coconut palms. It is known by various names in different regions and is common in various parts of Asia, Africa the Caribbean and South America.

Palm wine production by small holders and individual farmers may promote conservation as palm trees become a source of regular household income that may economically be worth more than the value of timber sold.

Palm wine is known as *emu*, *nkwu*, *oguro* in Nigeria; *nsamba* in the Democratic Republic of the Congo; *nsafufuo* in Ghana; *kallu* inSouth India; *Htan Yay* in Myanmar; *matango* in Cameroon; *tuak* in North Sumatra, Indonesia; *mnazi* in the Mijikenda languageof Kenya; *goribon* (Rungus) in Sabah, Borneo; and *tubâ* in the Philippines, Borneo and Mexico. In the Philippines, *tubâ* refers both to the freshly harvested, sweetish cloudy-white sap and the one with the red lauan-tree tan bark colorant. In Leyte, the red *tubâ* is aged with the tan bark for up to six months to two years, until it gets dark red and tapping its glass container gives a sound that does not suddenly stop. This type of *tubâ* is called *bahal* (for *tubâ* aged this way for up to six months) and *bahalina* (for *tubâ* aged thus for up to a year or more).*Toddy* is also consumed in Sri Lanka and Myanmar.

Tapping

The sap is extracted and collected by a tapper. Typically the sap is collected from the cut flower of the palm tree. A container is fastened to the flower stump to collect the sap. The white liquid that initially collects tends to be very sweet and non-alcoholic before it is fermented. An alternate method is the felling of the entire tree. Where this is practiced, a fire is sometimes lit at the cut end to facilitate the collection of sap.

Palm sap begins fermenting immediately after collection, due to natural yeasts in the pores of pot and air (often spurred by residual yeast left in the collecting container). Within two hours, fermentation yields an aromatic wine of up to 4% alcohol content, mildly intoxicating and sweet. The wine may be allowed to ferment longer, up to a day, to yield a stronger, more sour and acidic taste, which some people prefer. Longer fermentation produces vinegar instead of stronger wine.

Distilled

Palm wine may be distilled to create a stronger drink, which goes by different names depending on the region (e.g., *arrack*, *village gin*, *charayam*, and *country whiskey*). Throughout Nigeria, this is commonly called *ogogoro*. In parts of southern Ghana distilled palm wine is called *akpeteshi* or*burukutu*. In Togo and Benin it is called *sodabe*, in the Philippines it is called *lambanog*, while in Tunisia it is called *Lagmi* . In parts of Kenya (coast), it is known as "chang'aa". Chang'aa can be applied to wounds to stop heavy bleeding (mechanism of action not known). In Ivory Coast, it is called "koutoukou."

Africa

In Africa, the sap used to create palm wine is most often taken from wild datepalms such as the silver date palm (*Phoenix sylvestris*), thepalmyra, and the jaggery palm (*Caryota urens*), or from oil palm such as the African Oil Palm (*Elaeis guineense*) or from *Raffia palms,kithul* palms, or *nipa* palms. In part of central and western Democratic Republic of the Congo, palm wine is called *malafu*.

Palm wine tapping is mentioned in the novel *Things Fall Apart* by the Nigerian writer Chinua Achebe and is central to the plot of the novel *The Palm Wine Drinkard* by Nigerian author Amos Tutuola.

Palm wine plays an important role in many ceremonies in parts of Nigeria such as among the Igbo (or Ibo) peoples, and elsewhere in central and western Africa. Guests at weddings, birth celebrations, and funeral wakes are served generous quantities.

Palm wine is often infused with medicinal herbs to remedy a wide variety of physical complaints. As a token of respect to deceased ancestors, many drinking sessions begin with a small amount of palm wine spilled on the ground (*Kulosa malafu* in Kikongo ya Leta). Palm wine is enjoyed by men and women, although women usually drink it in less public venues.

In some parts of the Eastern Nigeria, the Igbo Land, palm wine is called "Nkwu Elu" or "Mmanya Ocha" (white drink). For instance, in "Urualla" and other "ideator" towns, it is used for traditional wedding. A young man who is going for the first introduction at his inlaws is required to come with palm wine. There are specific gallons of palm wine required depending on the custom of the various towns in some parts of the Igbo Land.

India

In India and South Asia, coconut palms and Palmyra palms such as the *Arecaceae* and *Borassus* are preferred. In southern Africa, palm wine (*ubusulu*) is produced in Maputaland, an area in the south of Mozambique between the Lobombo mountains and the Indian Ocean. It is mainly produced from the lala palm (Hyphaene coriacea) by cutting the stem and collecting the sap. In some areas of India, palm wine is evaporated to produce the unrefined sugar called jaggery.

In parts of India, the unfermented sap is called *neera* (*padaneer* in Tamil Nadu) and is refrigerated, stored and distributed by semi-government agencies.

A little lime is added to the sap to prevent it from fermenting. *Neera* is said to contain many nutrients including potash .

In India, palm wine or toddy is served as either *neera* or *padaneer* (a sweet, non-alcoholic beverage derived from fresh sap) or *kallu* (a sour beverage made from fermented sap, but not as strong as wine). Kallu is usually drunk soon after fermentation by the end of day, as it becomes more sour and acidic day by day. The drink, like vinegar in taste, is considered to have a short shelf life. However, it may be refrigerated to extend its life. Spices are added in order to brew the drink and give it its distinct taste.

In Karnataka, India, palm wine is usually available at toddy shops (known as *Kallu Kadai* in [Tamil], *Kalitha Gadang* in Tulu, *Kallu Dukanam* in Telugu, *Kallu Angadi* in Kannada or "Liquor Shop" in English). In Tamil Nadu, this beverage is currently banned, though the legality fluctuates with politics. In the absence of legal toddy, moonshine distillers of arrack often sell methanol-contaminated alcohol, which can have lethal consequences. To discourage this practice, authorities have pushed for inexpensive "Indian Made Foreign Liquor" (IMFL), much to the dismay of toddy tappers.

In the state of Andhra Pradesh (India), toddy is a popular drink in rural parts. The *kallu* is collected, distributed and sold by the people of a particular caste called Settibalija or Goud or Gamalla (Goundla).

It is a big business in the cities of those districts. In villages, people drink it every day after work.

There are two main types of *kallu* in Andhra Pradesh, namely *Thadi Kallu* (from Toddy Palmyra trees) and *Eetha Kallu* (from silver date palms). *Eetha Kallu* is very sweet and less intoxicating, whereas *Thati Kallu* is stronger (sweet in the morning, becoming sour to bitter-sour in the evening) and is highly intoxicating.

People enjoy *kallu* right at the trees where it is brought down. They drink out of leaves by holding them to their mouths while the Goud pours the *kallu* from the *binki* (*kallu* pot). There are different types of toddy (*kallu*) according to the season: 1. *poddathadu*, 2. *parpudthadu*, 3. *pandudthadu*, .

In the Indian state of Kerala, toddy is used in leavening (as a substitute for yeast) a local form of hopper called the "Vellayappam". Toddy is mixed with rice dough and left over night to aid in fermentation and expansion of the dough causing the dough to rise overnight, making the bread soft when prepared. In Kerala, toddy is sold under a licence issued by the excise department and it is an industry having more than 50,000 employees with a welfare board under the labour department. It is also used in the preparation of a soft variety of Sanna, which is famous in the parts of Karnataka and Goa in India.

Indonesia and Malaysia

Tuak is imbibed in Sumatra, Sulawesi, Borneo, Bali and parts of Malaysia such as Penang Island and East Malaysia). The beverage is a popular drink among the Ibans and other Dayaks of Sarawak during the Gawai festivals, weddings, hosting of guests and other special occasions. The Batak people of North Sumatra also consume palm wine. In Northern Sumatra the palm sap is mixed with raru bark to make Tuak. The brew is served at stalls along with snacks. The same word is used for other drinks in Indonesia, for example those made using fermented rice.

PALAKAPPAYANI (INDIAN TRUMFRUIT TREE)

Palakappayani is a night bloomer which has a strong stinky odour and attracts bats. It is an evergreen deciduous tree and is pollinated naturally by bats. The roots are sweet, astringent, bitter, refrigerant and are used for treating various health issues. The paste of the stem bark is applied for curing scabies and arthritis.

POOVARASU / CHEELANTHI (PORTIA TREE)

Poovarasu, which has beautiful yellow flower, is an evergreen tree with high medicinal value.

Its leaves, bark and flowers are used in Ayurvedic medicines for skin disease, jaundice and cough. Being salt resistant, this tree is more commonly seen in the coastal belt. Different parts of Poovarasu are used in the treatment of various disorders. It is also used for making furniture.

CHEMPAKAM (SAMPOO)

Chempakam is also known as champak and is a perennial flowering tree mostly found in Kerala. Its flowers are used to produce certain essential oils for aromatherapy.

Flower oil is used for cephalagia. Its strong fragrant yellow or white flowers are used in the perfume manufacturing. It is known for its medicinal properties and is used to cure stomachic, fever etc.

CHANDANAM (SANDAL WOOD)

Sandalwood is an evergreen tree known for its cooling and healing properties. It has small dull purple flowers in small bunches and round purple black succulent fruits.

Its sedative and cooling property makes a remedy for prickly heat, and inflammation problems of the skin. Sandalwood in various forms paste, dry powder or oil mixed with others can be used for profuse sweating. Sandalwood powder is useful in the treatment of gonorrhea.

Sandalwood oil is used in treating, painful and difficult urination and inflammation of the bladder. Incense from sandalwood has a calming effect and is conducive to pacify mind. Sandalwood paste is a vital ingredient in the Hindu rituals. It is believed that the sandalwood paste has the power to increase the concentration power of an individual when smeared on forehead constantly.

Sandalwood paste applied on head relieves headache and bring down the temperature in fever.

THEK (TEAK WOOD)

The teak tree is one of the tropical hardwood trees which can grow 30 to 40 m in height and often live to an age of 100 years. Its longevity is very great and specially noted for its capacity to withstand seasonal variations. The resin present in the heartwood in highly water resistant and it protects the teak from decay, insects and bacteria. Other characteristics of teak wood are durability, elasticity and presence of solid fibre. It is also used for making indoor and outdoor furniture making.

PUNNA (ALEXANDRIAN LAUREL)

Punna is a large evergreen tree, has hard and strong wood. It has been mainly used in construction of boats. Punnakka or the fruit is a favourite feed of the bats. It is a most wanted ingredient in cosmetics.

PLASU / CHAMATA (FLAME OF THE FOREST)

Chamata is the version of Sanskrit word 'Samidha', which is a wooden piece used for fire rituals. It is dry season deciduous tree and is used for timber, resin, fodder, medicine and dye. Its wood is water durable and is used for well curbs and water scoops.

EETI / VETI (EAST INDIAN ROSEWOOD)

Eeti is premium quality timber species internationally known as Indian Rosewood. It is used to make furniture and ornamental products. The tree attains a height of 20 to 40 m.

The colour of the heart wood darkens with time and the colour varies from light golden brown to light purple with dark streaks. The wood has high density and so is difficult to work on it and has fragrance. Rosewood has dimensional stability and therefore regains the shape after seasoning. The heartwood is durable and resistant to termites.

VENGA (MALABAR KENO)

Venga is a large deciduous tree which commonly grown in western and southern parts of India. Parts of the plant like heartwood, leaves and flowers have medicinal properties. It helps to improve insulin and proinsulin levels. The heartwood is used as astringent, anti inflammatory etc.

KAVUNGU (ARECA NUT TREE)

Kavungu is used as kodimaram during festivals in temples. Chewing the mixture of areca nut with betal leaf, tobacco and lime paste was an integral part of day to day life. Areca nut is an important item in dakshina to elders. It is a way to show respect to elders. The husk of the areca nut is green while fresh and becomes yellow to orange when it ripens. As it dries, the fruit inside hardens and can only be sliced using nut cracker.

KARIMPANA (PALMYRA PALM TREE)

The tree has a very important role in taking knowledge from past to the present generation with the inscriptions made on its leaves. In ancient times, manuscripts like Upanishads, Panchathantra were inscribed on Palm leaves. The tree is tall with black cylinder like stem and has fan shaped leaves. The leaves are used to make fans, mats, baskets etc. They can also be used for thatching roofs. The toddy obtained by tapping the tip of the inflorescence is a very popular beverage.

MANCHADI (RED SANDAL WOOD)

Manchadi is a medicinal plant and the seeds are bright red with hard shells. This plant is useful for nitrogen fixation. It is often cultivated as forage and an ornamental plant. The wood is extremely hard and is used in boat building and making furniture. The tree is useful in treating boils, used for making soap, tonic etc.

KUDAPPANA / TALIPPANA (TALIPOT PALM)

Kudappana which is one of the largest palm in the world has the largest inflorescence. The palm is monocarp, flowering only once, when it is 30 to 80 years old. The fruit takes a year to get matured and produces numerous round and yellow green fruit that has a single seed.

The plant dies after producing the fruit. In Malabar area, the palm leaves were used for making traditional umbrellas. The leaves are used for thatching and making hats etc.

CHEMPA / EZHACHEMPAKAM (TEMPLE TREE)

Chempa is a broadly grown tree having some very exotic followers. As it produces leaves and flowers even after being lifted from the soil the Buddhists think the tree as a symbol of immortality.

It is usually planted near temples and is used in temples for pooja. Funnel shaped flowers have five petals. This combined with sandalwood oil and camphor is best for itching.

VELLA MANDARAM (BUDDHIST BAUHINIA)

Vella mandaram, a perennial tree has sweetly scented white flowers. These flowers boom from January to April and then followed by long, slender seedpods. Before the leaves fall, the tree is covered with many delightfully fragrant, white, orchid shaped blossoms. The root of the plant is used as carminative and flower buds as laxative.

CHUVANNA MANDARAM (BAUHINIA PURPUREA)

Chuvanna manadaram is a large moderate sized deciduous tree that grows up to 10 m in height. It is related to peacock flower and is beautiful when in bloom. The flowers are five irregular, slightly overlapping petals in colours of magenta, lavender, purple etc. The bark, flowers and roots of the tree has medicinal properties.

PHANERA PURPUREA

Phanera purpurea is a species of flowering plant in the family Fabaceae, native to South China (which includes Hong Kong) andSoutheast Asia. Common names include orchid tree, Hong Kong orchid tree, purple bauhinia, camel's foot, butterfly tree, and Hawaiian orchid tree.

Description

Fig. *Phanera purpurea* flower (Kaniar) in Hyderabad, India.

Phanera purpurea is a small to medium-size deciduous tree growing to 17 feet (5.2 m) tall. The leaves are 10–20 centimetres (3.9–7.9 in) long and broad, rounded, and bilobed at the base and apex. The flowers are conspicuous, pink, and fragrant, with five petals. The fruit is a pod 30 centimetres (12 in) long, containing 12 to 16 seeds. Leaves are alternate.

Cultivation

In the United States of America, the tree grows in Hawaii, coastal California, southern Texas, and southwest Florida. *Bauhinia blakeana* is usually propagated by grafting it onto *P. purpurea* stems.

Biological activities

Phanera purpurea may possess antibacterial, antidiabetic, analgesic, anti-inflammatory, anti-diarrheal, anticancerous, nephroprotective, and thyroid hormone-regulating activity. Water extracts of the leaves of *Phanera purpurea* have been shown to have anti-ulcer activity in animals in the 'ethanol-induced gastric ulcer model'. Water extracts did not show any signs of toxicity when given to rats orally at doses up to 5000 mg/kg.

Chemistry

A wide range of chemical compounds have been isolated from Phanera purpurea including 5,6-dihydroxy-7-methoxyflavone 6-O-β-D-xylopyranoside, bis [3',4'-dihydroxy-6-methoxy-7,8-furano-5',6'-mono-methylalloxy]-5-C-5-biflavonyl and (4'-hydroxy-7-methyl 3-C-?-L-rhamnopyranosyl)-5-C-5-(4'-hydroxy-7-methyl-3-C-α-D-glucopyranosyl) bioflavonoid, bibenzyls, dibenzoxepins, mixture of phytol fatty esters, lutein, β-sitosterol, isoquercitin and astragalin.

Bibliography

Bagyaraj, D. Joseph: *Microbial Biotechnology for Sustainable Agriculture Horticulture and Forestry*, New India Publishing Agency, Delhi, 2011.

Bahati, J.: *Impact of Arboricidal Treatments on the Natural Regeneration of Species Composition in Budongo Forest*, Uganda. M. Sc., Makerere University, 1995.

Bawa, R. and P.K. Khosla : *Biodiversity of Forest Types A Community Forestry Approach* , BSMPS, Delhi, 1998.

Bhattacharya, Ajoy Kumar: *Community Participation and Sustainable Forest Development*, Concept, Delhi, 2001.

Bisht, Anand Singh; Krishan Datt Sharma and Matthew Prasad: *Food and Horti-Forestry Resources of Western Himalaya*, Jaya, Delhi, 2014.

Chakrabarty, Falguni: *Adaptation of the Santals to the Hill-Forest Environment*, APH, Delhi, 2012.

Falguni, C: *Adaptation of the Santals to the Hill-Forest Environment*, APH, Delhi, 2012.

Franklin B. Hough: *Elements of Forestry*, Discovery, Delhi, 1999.

Jackie C. Wooten : *Ethnomedicine and Drug Discovery*, Amsterdam, New York, Elsevier, 2002.

Jerram. M.R.K. : *A Text-Book on Forest Management : In Agriculture, Horticulture and Forestry*, Asiatic Pub, Delhi, 2006.

Kotwal, P.C. and Sujoy Banerjee : *Biodiversity Conservation in Managed Forests and Protected Areas*, Agrobios, Delhi, 2004

Kundu, S.S; O.P. Chaturvedi, J.C.Dagar and S.K. Sirohi : *Environment, Agroforestry and Livestock Management*, International Book Distributing Co., Delhi, 2008.

Miller, P. L.: *Some Dragonflies of the Budongo Forest*, Western Uganda. Opusc, 1993.

302 Handbook of Forest Ecology Musamali, P.: *The Ecology, Diversity, and Relative Abundance of* Small Mammals in Primary and Disturbed Forest Compartments *in the Budongo Forest Reserve*. . Sc., Makerere University, 1996.

Negi, S.S.: *Forest Soils*, International, Delhi, 2000.

Osmaston, A.E.: *A Forest Flora for Kumaon*, BSMPS, Delhi, 1994.

Prabhas Chandra Sinha : *Climate, Forest, Biodiversity and Desert*, SBS Pub, Delhi, 2006.

Puri, Sunil and Pankaj Panwar : *Agroforestry : Systems and Practices*, New India Pub, Delhi, 2007.

Reynolds, V.: *Budongo: A Forest and its Chimpanzees*, London, Methuen, 1965.

Sathe, T.V.: *A Textbook of Forest Entomology*, Daya, Delhi, 2009.

Sett, Rupnarayan: *Environmental Science : Botanical and Forestry Perspective*, Narendra Pub, Delhi, 2012.

Sharma, Ajay: *A Text Book of Forest Trees Entomology*, International Book Dist, Delhi, 2007.

Sharma, R.C. and J.N. Sharma : *Challenging Problems in Horticultural and Forest Pathology*, Indus, Delhi, 2005.

Solvang, R.: *Edge Effects on an Understorey Bird Community in Budongo Forest Reserve*, Uganda, Agricultural University of Norway, 1996.

Tewari, D D: *Management of Nontimber Forest Product Resources of India : An Analysis of Forest Development*, International Book Dist, Delhi, 2008 .

Praveen, T.: *Advances in Forestry Research in India*, Cyber Tech Pub, Delhi, 2009.

Ramappa, B.T.: *Some Economic Aspects of India : Agriculture, Forestry and Industry*, Serials Publications, Delhi, 2011.

Reynolds, F.: *Chimpanzees of the Budongo Forest*, New York, Rinehart and Winston, 1965.

Reynolds, V.: *Budongo: A Forest and its Chimpanzees*, London, Methuen, 1965.

Sathe, T.V.: *A Textbook of Forest Entomology*, Daya, Delhi, 2009.

Sheelwant Patel: *Indian Forests : Soil Water and Bio-Environment Conservation*, Pointer, Delhi, 2005.

Tewari, D D: *Management of Nontimber Forest Product Resources of India : An Analysis of Forest Development*, International Book Dist, Delhi, 2008

Index